FINDING MY FATHER'S FOOTSTEPS

Above: The United States 4th Infantry Division's 22nd Infantry Regiment loads at Plymouth, England, June 5, 1944, in preparation for landing on D-Day. They will be transported across the English Channel in this and other similar naval landing craft to Utah Beach, Normandy, France the morning of June 6, 1944.

The cover: The ruin of the Chateau de Fontenay, Saint-Marcouf, Normandy stands in silent remembrance of the battle the regiment fought here on June 10, 1944. It is one of the many battlefields that bear witness to the men, women, and children whose lives were forever changed by World War II.

Karen Marshall
Deeds not Words

FINDING MY FATHER'S FOOTSTEPS

A Journey Within the Battlefields of World War II

Karen Sisson Marshall

Deeds Publishing | Athens

Copyright © 2024 — Karen Sisson Marshall

ALL RIGHTS RESERVED — No part of this book may be reproduced in any form or by any electronic or mechanical means, including information storage and retrieval systems, without permission in writing from the authors, except by a reviewer who may quote brief passages in a review.

This is a work of nonfiction; all events took place. Where deemed appropriate, names have been changed to protect anonymity. The author has taken every effort to verify the accuracy of the information but assumes no responsibility for any errors or omissions.

Published by Deeds Publishing in Athens, GA
www.deedspublishing.com

Printed in The United States of America

Cover and interior design by Deeds Publishing

ISBN 978-1-961505-19-3

Books are available in quantity for promotional or premium use. For information, email info@deedspublishing.com.

First Edition, 2024

10 9 8 7 6 5 4 3 2 1

This book is dedicated to the memory of the girl in the red coat and to Sergeant Al Mead, who saved my father's life the night of the D-Day landing, June 6, 1944.

And to sons and daughters everywhere who have sought to better understand their parents and the experiences that defined their lives.

4TH INFANTRY DIVISION, 22ND INFANTRY REGIMENT IN EUROPE

LOCATION AND MOVEMENT DURING WORLD WAR II, 1944-1945

Contents

Text Notes	x
Map Illustrations	xi
Foreword	xiii
Prologue	xv
Part One: Death of a Soldier	1
Part Two: A Question of Honor	31
Part Three: Wading Into History	57
Part Four: One Soldier's Story	83
Part Five: In My Father's Footsteps	173
Part Six: A Deeper Understanding	227
Part Seven: Footsteps to Bavaria	267
Part Eight: The Death of a Soldier	371
Epilogue	388
Acknowledgments	393
About the Author	395
Abbreviations and Slang	397
Endnotes	401

Text Notes

My father's World War II narrative is printed for the first time in this book. Bringing life to a voice that is silenced by death is a logistical as well as emotional challenge. Since both my father and I shared a love of typing, it seemed a natural decision to give his voice a `font reminiscent of a manual typewriter` while the rest of the book was a standard font. To help the reader, a list of the slang and abbreviations that he used in his journal is also provided at the end of the book for reference.

Map Illustrations

BY ANN M. BEDRICK

Since this is a story that unfolds on the World War II battlefields of Europe, Ann and I developed a series of illustrated maps to help guide the reader on the journey. These are guides only to where the men were speaking and fighting from during their long march from Normandy, France to Bavaria, Germany. There is no intent to capture the enormous war effort that was waged all around them or to provide precise coordinates or strategy.

LEGEND

Foreword

Robert O. "Bob" Babcock
4th Infantry Division Association,
Historian and President Emeritus
22nd Infantry Regiment Society, President Emeritus

I never tire of reading stories from and about my beloved 4th Infantry Division. World War II history has fascinated me ever since I was a young boy and is stronger now than it was then. As we commemorate the 80th anniversary of the D-Day landing on June 6, 1944 (in 2024), we have another first-hand account of that significant day and the following 11 months in world history. Part of the story was written by Lieutenant John Sisson, an Infantry rifle platoon leader, as he was home in January 1945 as a twice wounded veteran participating in a war bonds drive. It wasn't until after his death that his daughter found what her father had written so long ago.

Most interesting among Lt. Sisson's souvenirs brought home at the end of the war was a copy of Adolf Hitler's book, Mein Kampf, signed and annotated by Hitler's top supporter, Heinrich Himmler. That started Karen Sisson Marshall on her journey to research how her father came upon the book as he continued the fight across Europe after returning from his 30-day leave home.

An educated historian herself, Karen spent months and years researching her father's time in Europe in World War II, including a trip to Normandy with her daughter tracing his footsteps from the D-Day landing.

I am proud to have been both a historical advisor and the publisher of this great addition to World War II, 4th Infantry Division, and 22nd Infantry Regiment history.

Steadfast & Loyal—Deeds not Words!
—Bob Babcock—April 1, 2024

Note: For impatient historians, the first 12 chapters of the book set the stage before Karen and her father walk you through the 22nd Infantry Regiment's fight from D-Day to VE Day in the remainder of the book.

Prologue

"As you go about your day this June 6 and think of the duties you have and the reality you face, take a moment to remember the ordinary men who did extraordinary things to make our world ordinary. God bless the men of D-Day."

— **Vietnam Veteran,**
4th Infantry Division Association Newsletter, 2020[1]

Some journeys begin before we are born. This one started with a break in a storm on the morning of June 6, 1944, off the coast of Normandy, France. The ominous seas and skies had teemed with thousands of planes and ships since midnight. On the great, churning, gray English Channel, men, boys many of them, from every state in America had merged into a mighty Allied Army with their comrades from England, Canada, Europe, and Australia.

The 3,000 Soldiers of the 22nd Infantry Regiment, part of the United States 4th Infantry Division, began landing at 6:30 a.m. on Utah Beach, one of the five beaches chosen by the Allies for the invasion of Western Europe. My father was a second lieutenant with the 22nd Infantry Regiment's Second Battalion. His regiment suffered minimal casualties that day, thanks in large part to the terrible price paid by the men of the 82nd and 101st

Airborne Divisions, who had dropped from the sky in the early morning hours to assist in securing the beachheads of Normandy, particularly at Utah Beach.

By September 11, 1944, the men of the 22nd Infantry Regiment had fought without pause across France and Belgium and were the first Allied unit to penetrate Germany's vaunted Siegfried Line. On November 14, 1944, they faced a miserably cold and foreboding sea of hills and towering trees in Germany's Hürtgen Forest. In the ensuing brutal eighteen-day battle, the regiment would sustain 2,678 casualties: 83 percent of the unit's authorized strength. My father was one of them.

Replacements and returning wounded filled gaps, but by the war's end on May 8, 1945, the 22nd Infantry Regiment had suffered casualties nearly three times its authorized strength. From the Regimental Headquarters down, no unit had maintained its full complement of officers, men, or equipment in eleven months of combat. Only a small percentage of the men of D-Day remained in the field when the battered 22nd Infantry Regiment assumed peacekeeping responsibilities in Dinkelsbühl, Germany.

Fifty-five years later, William C. Montgomery, 22nd Infantry Regiment medic, paid homage to all the infantry soldiers who fought for freedom in Europe from June 6, 1944 to May 8, 1945. Pushing through snow drifts up to his waist in the steep mountainous terrain of the Hürtgen Forest, he was part of three four-man stretcher squads trying to rescue two men who had spent the night wounded and alone. Montgomery described what happened:

> One man had frozen, poor guy. The other was a real Survivor—with a capital "S."
>
> His leg was shattered, but he had made a tourniquet of his belt.

He burrowed down into the snow and covered himself with his shelter-half, lit an alcohol cube with his Zippo lighter, and shoved his rifle butt into the small flame. When the fumes got to him, he put out the flame, flapped the shelter-half and settled down for a while, then started over again.

We strapped him securely onto a litter and began a rough mile-long litter haul, with one squad breaking trail in the snow, one carrying, and one following along catching its collective breath. We wanted to make sure this survivor got to the aid station and eventually home.

I remember him well because he was such a vivid example of the hell riflemen had to live through every day. Even now, fifty-five years later, I'm still in awe of what guys like him did. My squad and I carried a lot of them off the line—enough to know it was the infantry who won the war.[2]

Part One

THE DEATH OF A SOLDIER

"…perhaps more men should know the expense of war, for it is neither a fit way to live nor die."[3]
— **Charles Wertenbaker,** *Invasion,* 1944

Lt. John F. Sisson with his father, Warren B. Sisson, Winter 1945[4]

1. Ponderings

My father, John Fletcher Sisson, had four war stories he told—and his children were chuckling at the end of each one. They were all in reference to World War II, the largest war in the history of humanity, when 50 - 75 million people—about three percent of the world's population—perished. This does not count lives ruined just lost. A war that tore nations and families apart and ended with the atomic bomb, the most devastating instrument of war that had yet been devised. The war that would define the modern world—the world we live in today.

The thought that lighthearted stories about such a war might be inappropriate never crossed my mind growing up. When it finally did, I discovered most of my contemporaries found themselves in the same situation. Although all our parents were affected by the worldwide conflict, the great majority of them never spoke about it, unless casually like my dad. And yet, from the start, my father and his friends recognized the seriousness of the situation.

On September 1, 1939, one of my father's fraternity brothers at Ohio State University said what many were thinking about the turmoil that would follow Germany's invasion of Poland.

Dear John:

I have just thought of a possibility. What if Poland's traditional enemy, Russia, would attack Poland on the eastern front? This would remove part of the guilt from the hands of Hitler and force the English and French to have a conflict with Russia because of the guarantee France and England have given Poland.

After listening to the latest reports from abroad concerning the international situation I hold out no hope that war can be averted ... It's like watching a great tide approach bringing on chaos. I hesitate to give Hitler all the credit for the ensuing calamity. It seems too big to be the plan of one man. It is the biggest physical thing that could happen on a little planet like the earth. It is hardly less spectacular than seeing the sun go out like an incandescent light.

One thing for certain is that you and I will be performers in the biggest show on earth one of these days.

I hope I am strong.
Blankity kai blank[5]

Dad's journal entry on October 13, a month later, confirmed his friend's forebodings:

```
We have a war in Europe now—since Sep-
tember 2nd in fact. Maybe this the first
Friday the 13th of the overseas mess will
be the unlucky day. Yesterday Chamberlain
told Parliament and the world that Hit-
```

ler is a low one—a liar that the German's
word can't be trusted. They're having a
great time. Hitler speaking for the Axis,
Chamberlain for the Allies—both determined yet both trying to push the technical blame for this war on the other. God
knows the reason is clear enough—Fascism
v. Democracy—but the statesman must be
conscious of his biography I suppose.[6]

When my father wrote, he did not know there was something far more fearsome than fascism at work deep within Adolf Hitler's Third Reich. It was the planned annihilation of "impure races."

One of the stories my father told concerned a faded red book that was always shelved at random in our home. The book was an artifact from the war, booty. It was a copy of Adolf Hitler's *Mein Kampf*, annotated by Heinrich Himmler. Dad kept a thin paperback book next to it that was entitled simply *Himmler*. The front cover featured a weasel of a person. On the back cover, written in bold black letters, it said:

> Hitler was surrounded by sycophants and toadies, men who competed for shares in the spoils of power. Goring, Goebbels, and Heinrich Himmler stood out among this new German aristocracy, and of the three, it was Himmler, the obsessive gatherer and filer of information, who eventually came to wield most real power. As head of the Gestapo and the concentration camps, it was he who organized the extermination of more than 10,000,000 people.[7]

Documents from the Yad Vashem Archives in Jerusalem reveal that on June 22, 1944, Heinrich Himmler reflected on the Final Solution he had devised, inspired in part by what he had read in that faded red copy of *Mein Kampf* now owned by my father:

> It became necessary to resolve another important question. It was the most terrible task and the most terrible order which could have been given to any organization: the order to solve the Jewish problem...This is a good time—we had the toughness to exterminate the Jews in our sphere. Don't ask how difficult it was. As soldiers, try to understand how difficult it is to execute such an order. The order was necessary.
>
> I told them—First, there is the order, and second, our conscience tells us to carry on ruthlessly with this cleansing process. And if anyone comes along and says, "Well, you know, I understand very well your killing grown-up Jews, but how can you kill women and children?" I say this: "These children will grow up one day..."[8]

A faded red book.

This book represents the worst of humanity. The eyes that read its words, the soul that processed their meaning and the hands that underlined passages oversaw the brutal organized murder of millions of men, women, and children. It was the evil heart of the war in Europe that engulfed my father and his peers.

As a historic preservationist, I have handled many documents and photographs and assessed structures and landscapes. Artifacts, however, are not something that I deal with on a regular basis. I know that buildings can tell many stories of our shared heritage when one considers the minds and hands that built them and the lives that worked and lived within their walls. But what could a faded red book once owned by Heinrich Himmler, the man who uttered those words, say to me?

Or to my father who took and kept it? My father, who on November 24, 1939, the day after his 22nd birthday, wrote in his journal:

```
To be steadfast in friendship, to be
steadfast to self, to be steadfast in
mind.
```

2. Shock

FEBRUARY 28, 1991

My breath was coming in erratic bursts that condensed in the chill of the February morning. I had received the telephone call while at work. My aunt's voice had been measured. "Karen, I am so very sorry, your father died this morning."

My co-workers urged me to stay with them, but I wanted to be alone with my panic and my tears despite the cold morning. As I struggled home, walking past the familiar brick row houses of my neighborhood, I experienced the strange sensation of moving in an alternative universe. Around me life buzzed, chirped, and whirred as if unaware that the very essence of existence had altered. Dizzy, I sat down on the nearest front stoop.

My neighbor Louise found me. "Are you alright?" She sat next to me, a frown on her face.

"No," I whimpered. The taste of the word and my tears surprised me. My head instinctively found her shoulder. I somehow stammered, "My father died." The tears kept coming.

"Oh no." Her searching eyes softened, and she whispered, "Not John. Oh my, oh my." She put her arm around me. We sat silently, rocking together.

I broke the silence, "It was sudden. My mother found him." Louise hugged me closer and smoothed my hair. A passerby looked at us with concern and she faintly shook her head. "Come home with me," she urged. When I didn't answer she asked, "Do you want company?"

Somewhere in those words I found myself again. "Oh Louise, you saved me. Thank you, thank you!" We started laughing, then crying.

"What a handsome man, what a beautiful couple," her eyes glistened with memories from our parties which always included my parents when they visited from Florida.

We stood and walked arm in arm toward my long narrow brick rowhouse. I hugged her again and went inside.

My husband Gary and I had moved to Federal Hill in Baltimore, Maryland in 1989. Earlier that decade, the South and I had developed a friendship, if not a brief love affair. Gary managed the development of independent television stations for license owners, and we wandered for his work from the mountains of Asheville, North Carolina to the savanna of Charleston, South Carolina. The local stations he helped to create were beset by rapid industry changes, and when the station in Charleston faltered, we left our home on Tradd Street for a new project in Baltimore.

Otto, our dog, ran to welcome me, then paused and began to whimper. She climbed up next to me as I collapsed on the sofa. Otto, our joy, our girl named Otto. No one looked too carefully after I helplessly accepted "him" from two youngsters who were giving "him" away. We named the small, fuzzy, black puppy after the neighborhood he came from, Otterbein. Finally, an old friend declared one day after dinner, "If that dog is a male then I am a Leprechaun!" We changed her pronoun but not her name.

Otto had already seen me through two heartbreaking miscarriages. I held her tightly and prayed she could help me endure this new unbearable loss. During the spring our family was shocked by my father's diagnosis of a rare form of cancer. It had wrapped around his femur, requiring a complex and delicate operation to avoid the amputation of his leg. He faced this news with the normal outward calm he always used to address issues of great concern. Perhaps this was to be expected from a man who carried a Nazi bullet in his abdomen for his entire adult life.

I had visited my father that summer at the cancer center in Gainesville, Florida where he spent weeks receiving intense radiation prior to the difficult surgery to remove the tumor from his leg. Somehow, he maintained the gentle humor he had developed in retirement. When I was growing up, he had firm views on life, which he expressed with a brusque, decisive delivery. Stern, almost abrupt, he terrified my friends and enraged me. This confident cadence, developed during a legal career that included arguing a case before the Supreme Court, miraculously softened with his retirement.

There was another change. He discarded the ever-present pipe. Pipes were such a fixture in our lives that he once set his jacket on fire lighting one on a ski lift. His decision came after a week at the center. Riding the elevators with people who had lost portions of their faces and jaws to surgery for smoking-related cancers clinched the deal.

My father once told me that despite his war record, it was my mother who was brave, not him. We were talking about her battle with breast cancer, which had led to two mastectomies ten years apart. The nine months leading up to his untimely death at 74 proved otherwise. They were both cancer warriors, scarred and battered, yet bravely following orders, never yielding.

The phone rang and startled me from my thoughts. It was my mother. She recounted what she knew in a slow and measured voice. It had been a morning like any other except that when she emerged from their bedroom, she noticed something odd. The coffee pot was not turned on. She realized she hadn't talked to her husband yet that morning. She walked to his study, opened the door to his bathroom, and found him dead on the floor. Her voice broke, and she stopped speaking.

"It's all so terrible Mommy. We will be there tomorrow."

"Oh Karen." She began to cry.

I don't remember what we said next. I sometimes wonder if our minds protect us by forgetting the unbearable. I know I told her I loved her.

I later learned that after my mother found my father she ran screaming into the driveway where her neighbors, by some act of Providence, were getting into their car. She never lost consciousness, but she never fully recovered from the shock. Bloody handprints could be traced down the wall of his bathroom. Dad had somehow composed himself after the aneurysm erupted from the deep surgical wound in his leg. He was lying on his back with his hands folded on his chest.

A haze of telephone calls and visitors crowding our living room somehow followed. My mother's first cousin and dear friend protected me from small talk. Gary refilled drinks and Louise magically appeared with food. Finally, just the immediate family remained. Someone said, "Tell us the war stories, Karen." I was hesitant, but I felt less alone as I recounted the familiar tales, even though the stories sounded a bit like a European holiday.

Dad was an infantry lieutenant and landed on Utah Beach on D-Day. His first major battle was at an unnamed chateau where his commanding officer and many comrades were killed. The details were vague, but we knew he had been shot and the bullet that had lodged in his abdomen remained with him for the rest of his life. Somehow, he walked to the aid station with other injured men and when the medics finally looked at him, one said, "Man, what are you doing alive?"

The next story involved his return to the front lines and a hotel in Belgium where he negotiated warm baths for himself and his buddies. The only reference to war was the cold and mud and the fact that somewhere along the line he took a second bullet through his hand. He would tell us with a laugh that he was teased to no end because soldiers were accused of putting their hands up in fox holes hoping for an injury and relief from combat.

Everyone was smiling through their tears as I ended with the script Ohio saga.

Born and raised in Columbus, Ohio, my father was an Ohio State Buckeye through and through. Even though we left Ohio for the East Coast when I was in second grade, I remember that in 1958 you could lie down in the street and take a nap during a home football game. Nothing moved but the corn in the fields. Buckeyes ruled. And one of the most famous Buckeye pregame rituals was performed by the Ohio State Marching Band in their "Script Ohio" routine. The band wrote out Ohio while playing a rousing marching tune. Few fans knew the music was "Le Regiment de Sambre et Meuse," a patriotic French march penned in the 1870s.

Now, it's 1944 and somewhere in a bar in war-torn France. Lt. John Sisson suddenly hears "Le Regiment de Sambre et Meuse" being played. Astounded the French know this Ohio State marching song he decides France must indeed be a country worth liberating and begins to execute the "Script Ohio" routine with several other GIs. The Frenchmen in the bar, overwhelmed that the Americans recognized their patriotic march, fall in behind them. Soon the entire bar is executing the "Script Ohio," no doubt flawlessly. When the march ended, I am sure there were smiles and congratulations in broken English and French as well as toasts.

This was the entire World War II narrative that my family knew, except for one last story that for some reason I did not tell that evening: How my father came home from the war with Heinrich Himmler's personal copy of *Mein Kampf.*

After tearful farewells, I continued to stand on the worn brick sidewalk in front of our home to watch Gary take Otto for her final walk of the day. It was probably nearing 11 P.M. My long black wool coat barely kept the chill from driving me back inside.

As I watched their progress, I suddenly smelled summer and heard the rush of wind in fully leafed trees. It was 1967 and I was blissfully seventeen. Sitting in my boyfriend's yellow GTO while the seniors planned the night's party, I leaned on my firm tanned arm smelling the mixture of body lotion and baby oil from my day at the beach. Vaguely aware of the activity around me, humming a tune I don't remember, I suddenly felt at perfect peace with the world.

A chilling breeze tickled my ankles, and winter jerked me back to the day's cold realities. I pulled my coat closer, surprised that this old memory had decided to surface now. Even today I am startled by how clearly that summer afternoon remains fixed in my memory. It stands in such a stark contrast to the angry and confused time that began two years later when I ran away from home and friends. How did I fall so far out of harmony with life after that moment of almost perfect bliss?

After many years, I finally concluded that far from the obvious — a young girl at the top of her game — my moment of spiritual harmony had much more to do with my family than high school popularity. My father's disciplined yet generous approach to life resulted in a home that was, in a word, secure. Imperfect like all homes, but a place one could return to at any time for welcome and refreshment. I wondered if it was that safety and resulting sense of well-being that inadvertently opened the door to God's grace that lazy afternoon.

I considered that moment one last time after I fell into bed, exhausted from too much emotional pain and liquor. I finally realized why I had revisited that summer day. Our haven. My father's gift to his family. Now snatched away forever.

I had crashed headlong into the wall of silence that death immediately erects. I did not know I would one day embark on a journey to find a way around that wall.

3. Mysteries and Secrets

MARCH 1991

Gary, Otto and I drove to Florida the next day. My parent's home was in a gated golf community called Sawgrass in Ponte Vedra Beach, Florida. Upon arriving, I tried to calm my pounding heart and went inside.

I found my mother sleeping with her back towards me on the beige and lavender flowered bedspread that epitomized the cheerful elegance of my parents' bedroom. Their wedding album was open beside her. Quietly, I sat down and looked at the pictures. Her formal wedding portrait captured her regal posture and shy smile. I had worn the same elegant dress and heirloom veil when I married Steven, my first husband. We were living a bohemian life and selling motorcycles. In retrospect, it seems an odd choice.

The yellowed, wrinkled newspaper articles from the December 4, 1948, wedding reported:

> The bride is a graduate of Chevy Chase Junior College at Washington, D.C. Mr. Sisson graduated from Ohio State University and the Law School at the University of Michigan. He now practices law with…[9]

I became immersed in the pictures of their old friends and

my grandparents and touched their faces gently with my finger. With a few additions in Darien, these were the people who had provided the foundation for my life growing up. Successful, attractive, deeply conservative, fun loving and smart.

My parents were much younger than my 41 years; she was 28 and he 33. Their courtship was brief; Mom had waited for the war to end to begin her life. We had been told that when Dad failed to propose within a year, Mom announced she was planning an extended visit to help her cousin with a new baby. When asked when she would return, she told Dad, "That will depend on you." They married soon after.

His best friends, Dick and Marjorie Taylor, had introduced them. Sometime in early 1947, they invited my father to a dance at Fort Hays in Columbus, Ohio where Marjorie's father was the commanding officer. Lieutenant Colonel Lyons and his wife were visiting with their daughter, who planned to attend the same dance. A charming and vivacious blonde, Marjorie was drawn to the beautiful dark-haired daughter, and decided to play matchmaker with Dick's help.

The story goes that when his friends suggested they leave, my father hesitated and said, "I think I will have one more dance with the Dragon Lady," a reference to an exotic Asian character from the comic strip *Terry and the Pirates*. When Dad told the story he would laugh and say something like, "It was really your mother's legs." We nodded in appreciation. Political correctness had to do with being a member of the Republican Party—the correct party—and not much else.

One of the songs my mother liked to sing was Frank Sinatra's "The Last Dance." It always evoked that romantic beginning for me. In 2013, their granddaughter would sing it for her as we said our final farewells to John and Betts—as only my father called his wife.

My mother stirred. I still didn't want to wake her. After a time, she whispered, "Karen?"

"Yes, Mommy."

She sighed, turned toward me and whispered, "I was supposed to die first."

I knew she meant it. I took her hands and we both began to cry.

I don't recall when David and Elizabeth, my brother and sister, arrived at the house. I don't remember what food we ate or what we talked about.

One of the few clear memories I have from those first few days is standing in the funeral home next to my father's body with David, Gary and the undertaker. My father lay almost peacefully, a sheet draped over him just as artists have depicted Christ in his tomb. He was a religious man, and I felt he would have understood the image. I was surprised I felt comfort in sharing death with him. How I missed him.

The undertaker asked if we would like a lock of Dad's hair. My brother and I said yes. And then something happened that I have pondered ever since. In words that I could understand with absolute clarity even though there was no actual sound in the room, my father said,

It's OK to die.

I was so startled. It wasn't until later that afternoon that I reflected on that final visit. As the gentle winter sun filtered through the large ficus tree that stood in the corner of my parents' living room, my main question was, why me? No one else present heard anything. Over the years, people have shared their similar experiences. The voice is clear yet unheard by the ear. It's as if one's soul processes the words.

I couldn't help but wonder if Dad was trying one last time to help me to not be so afraid of death. I have an anxious nature, and in my youth, I developed the habit of feeling for my pulse. This drove my father crazy, especially when I was young and indiscreet. "Believe me," he would say, "if your heart stops you will know it." He made various attempts to quiet my fears. This included the gift of a framed copy of Ecclesiastics 3, which begins "To everything there is a season, and a time to every purpose under heaven: A time to be born, and a time to die." It now hangs in my daughter's room.

I began to fuss with a small burn hole in my chair's fabric. My mother had upholstered their armchairs in a dark brown fabric to hide all the burn holes from Dad's pipes even though the rest of the living room reflected a brighter Floridian theme. I affectionately smoothed it down, then leaned over to give Otto a hug. Mom's voice interrupted my reflections. She was asking if I would look through my father's papers. The bankers were coming the next day to review the terms of the trust he had established for her.

I found it unsettling to sit at my father's desk in his study, but I couldn't help being amused by his Depression-era thriftiness. It was clearly an old surplus desk, probably from the basement at the Columbia Gas corporate headquarters in Wilmington, Delaware, where he spent the final years of his career. A free metal office desk in good working order was vastly preferable to a purchased wood one. One of his favorite sayings was, "I shop at Sears so your mother can shop at Bloomingdale's."

A framed picture of his Beta Theta Pi fraternity house at Ohio State University hung on the wall above his desk. I glanced around the room at familiar objects—a lamp with a hunting scene, his father's humidor and of course his collection of pipes on the chest of drawers my mother had antiqued in Darien. I returned to the matter at hand. His files were orderly and clearly labeled. I didn't need to look at the contents of each file to tell Mom the bankers would have no trouble reviewing his accounts.

My next destination was the unfinished bedroom on the second floor which served as the attic. The room had an inviting feeling despite the random placement of several framed posters, an old lamp, several upholstered chairs and perhaps a dozen book boxes. My mother's scrapbooks were arranged against one wall on two card tables. My father's large metal file cabinet stood near the center of the room.

I sat down in a chair and opened the 1986 scrapbook from my parents' trip to Italy. I had forgotten that their adventure had overlapped with the nuclear meltdown at Chernobyl. I had called my parents to ask if they were coming home, and my mother laughed, "I waited too long for this trip. Don't worry." I guess nuclear meltdowns were small trouble compared to the Great Depression, World War II and the Cuban Missile Crisis.

My next choice was a 1964 scrapbook from the Darien years. I became lost in the array of Playbills, pictures, and invitations. A ski weekend with their Darien friends caused me to laugh out loud. There was my mother, who never skied, looking just like she had stepped out of the après-ski scene in *The Pink Panther* while we, looking flushed from the slopes, wore the slightly sagging sweaters she had knitted. Dad led the charge on family trips, water skiing and beach picnics, New York City Christmas lights and ice cream after church while she gently encouraged each of us from a poised and slightly removed vantage point.

I wiped away a tear. I just wanted to sit there, looking at them, at us, smiling together in Connecticut. To reach back behind the door now closed, before the war in Vietnam, before the move to Wilmington and before cancer began taking its toll on our family. I finally returned to the matter at hand, my father's files.

The lower drawer in the cabinet was filled with several scrapbooks from high school and college and paperwork from various real estate sales and investments. I found Dad's diary from the grand tour of Europe he took in 1935 with his first cousin and favorite aunt. I loved his stories from that trip, which he always told with great pleasure, ending with, "I slept in Katharine Hepburn's bed, but unfortunately she wasn't there." I placed the diary and his scrapbooks with my mother's albums.

I opened the top drawer of the cabinet with difficulty as it groaned from disuse. I almost closed it again when I found more miscellaneous files but paused when I noticed several books. I removed a hardbound dark blue volume for a look. *4th Infantry Division, 22nd Infantry Regiment.* I opened it and scrutinized a map that illustrated the "Advance of the 4th Infantry Division toward Germany in 1944."

The familiar faded red book was next, along with its accompanying paperback. So, this is where Heinrich Himmler's copy of Adolf Hitler's *Mein Kampf* ended up after this move. We all knew the story.

My curiosity piqued; I now examined each file with care. A weathered cardboard mailing envelope with the familiar red stamp "Photographs Do Not Bend" was from Dale Laboratories—my father's preferred film developer. He always kept his Naugahyde case with his camera and lenses close by his side. I opened the envelope and read the copy of a letter that preceded several photographs.

FINDING MY FATHER'S FOOTSTEPS

February 19, 1980

Dear Bill,

Last October, I had a chance to return to the Cherbourg Peninsula where I tried to retrace some of our steps in 1944. In a rented car, I drove from Cherbourg and was able to find Montebourg with no trouble. Several years ago, I ran across some 1:50,000 maps of the area where we landed. These helped tremendously.

The map doesn't show Utah Beach itself, but as indicated, it is about 2 miles below the bottom of the map. I marked in what I recall as our general line of march after landing. I had thought that the first village our platoon came to after going up a road inland and parallel to the beach, was St. Martin de Varreville, but somehow it didn't fit my memory when I was on the ground.

In any event, the main points were Azeville and the Chateau de Fontenay, which I did indeed find, together with nostalgia and melancholy...[10]

I stopped reading as he began to describe in detail the photographs that I took out of the box, one by one. I put the letter and pictures carefully aside. The next file folder was full of yellowed typed pages. I removed the first sheet of paper and read it.

```
This is Saturday night, the sixth of January
1945. Just about a year ago this story, or
whatever it can be called, started. Since ev-
ery second letter from Aunt Kathryn tells me I
ought to keep a record of this period, I'll say
it's to satisfy her—but really it is a pretty
dull Saturday night, and if Sunday morning turns
out better, this will undoubtedly never be fin-
ished.[11]
```

My father's voice spoke out from beyond the grave. It was a narrative from his war years. Spellbound, I looked at the manuscript, the pictures and the letter.

Why did I just listen to his war stories and never ask any questions? What was wrong with me? Was life, in the end, one great regret for opportunities lost, stories not heard, questions not asked? Who was William Weingart? My mother would know, but why hadn't they shared this with us? We were all proud of Dad's two Purple Hearts and military service in World War II and the Korean War.

I called downstairs, "Mom, did you know that Daddy wrote a story about his war experiences?"

"Can you come down here? I can't understand you."

I gathered up the documents and hurried down.

"Did you know that in 1945 Dad wrote a journal about his D-Day landing and that he returned to Normandy in 1979?"

She took the documents from me and looked at them with a puzzled frown. "No."

I was speechless for a minute. "What do you mean?"

She looked so tired; it broke my heart. "I had no idea this

existed. I have some vague memory of him telling me he went to France during one of his business trips to London."

"These were in with a bunch of military papers and books. It looks like he went to reunions; there are programs. Do you want me to bring the folders downstairs?"

My mother shook her head, "I know he was very active with the division reunions and kept in touch with a number of men. I met several."

She walked over to the dining room table. Sitting at her place she began to read the worn typed pages. I sat down at my place and looked toward the empty chair at the head of the table. It was at this table, in our dining room in Darien, where I had shouted my disagreement with my father's stand on the war in Vietnam from late 1968 until the end of the Nixon presidency.

The Viet Cong and North Vietnam launched the Tet Offensive in South Vietnam in 1968 as I was turning eighteen. By Thanksgiving of that year, I was a "dues paying member" of the counterculture and anti-war movement. Most dinners you could cut the tension with a knife, although I was the only one who raised my voice. Sometimes my father's silence was harder to take than the facts he countered me with at every turn. My mother, brother and sister would stare silently into their food as we clashed. I usually left the table in tears shouting, "You just don't understand. People are dying, you just don't care!!"

My mother seemed lost in thought. I picked up the letter to the unknown William Weingart and reread it. So, Dad was wounded and several of his comrades killed at the Chateau de Fontenay. The battleground was no longer a nameless place in France. That clarified one story. I wondered if the monument he had mentioned in the letter ever got erected.

It wasn't until Mom went to bed that I finally had an opportunity to read the manuscript I had found in the attic. I stopped after Dad landed at Utah Beach on D-Day. I had a difficult time with all the unfamiliar military terms, and I could not stop asking myself,

"Why? Why didn't he share the story with his family!"

Unbeknownst to me, I had been swept up into the journey that had begun all those years ago on a beach in Normandy, France.

4. Endings and Beginnings

1993

On the second anniversary of my father's death, I was sitting on a yellow loveseat in my living room in Chester County, Pennsylvania, thinking back on all that had transpired since he left us.

The Florida morning of Dad's service in 1991 was everything a northerner could pray for, with a gentle breeze off the ocean and soft sunlight lifting our spirits despite our heavy hearts. One of his golf buddies had contacted me and insisted we needed to hold a service even though our family was planning a formal memorial that fall in Darien, Connecticut. Dad was well liked and respected among the men. They wanted to honor a fallen comrade.

I was barefoot for most of the service, feeling the cool sand against my feet, listening to the waves on the shore and the cry of the gulls overhead. A simple tent had been erected in front of the beach club, a table with a bible, cross and humidor created the sanctuary. My mother had been unable to decide where to put Dad's ashes, and my brother suggested the humidor, which always occupied an important place in his room. Forty men and a few wives joined our family in quiet remembrance.

We had learned another of my father's secrets in the days following his death. When the surgeon called to express his sorrow, he had revealed the truth my father alone had been living with when he said, "It is, in many ways, a better way to die than from lung cancer."

Startled, my mother asked him what he meant.

"Then he didn't tell you?"

"Tell me what?"

"The cancer metastasized. He had lung cancer."

Mom never revealed the shock, sadness, or anger this statement must have elicited. She just told the family what the surgeon had said one night when we gathered after dinner.

Sitting in the home Mom and I now shared together, I painfully recalled haranguing her about not adequately feeding him. His gray complexion and weight loss haunted me every time I visited Florida during the five months following the surgery.

Reverend John Dolaghan, first pastor of Sawgrass Chapel, officiated at our oceanside service. A decorated Vietnam Navy veteran, he told us how much he enjoyed talking after church service with Dad about World War II and Korea. "There is no difference," he said to my mother, "between losing someone suddenly or slowly except for one thing. When there is a sudden death, people experience shock. No one is ever ready or prepared for the death of a loved one." She repeated those words to us with regularity, no doubt clinging to them for strength.

I knew we would like Reverend Dolaghan. My father picked his churches based on the message he heard from their ministers. I remembered our last service together. It was my church, the First Scots Presbyterian Church in Charleston, South Carolina. I was proud to share this special place with him. I loved the look of amused yet thoughtful irony which always assured me he was enjoying himself. He looked dapper in his madras jacket, pockets filled with pipes, and his straw hat on the pew. My mother sat beside him, fashionably dressed.

I would return there with that memory to christen the granddaughter he never met, and I would name her after his Aunt Kathryn, who had been such a remarkable force in his life. Aunt Kathryn's father, Fletcher Marion Sisson, was a prairie Methodist minister—bravely and faithfully traveling the untamed West in the mid-1800s. As family matriarch, Kathryn assembled her flock with the same faith in God, becoming the first Dean of Women at Ohio Wesleyan University, then evolving into the wife of a financial baron. Dad was very special to her. Much of his faith came from her example and, in turn, they had set an example for my relationship with God and an ethical life.

My sister read a letter from Kathryn's son, Ellis Phillips, at the service. It began:

> Reminiscences of John Fletcher Sisson, Named for our grandfather, Reverend Fletcher Marion Sisson:
>
> John is my first cousin, my mother's nephew, and I should know him pretty well. I do, in many ways. But we lived in different parts of the country and our paths crisscrossed but irregularly. John was about three and a half years older than I and I looked up to him literally, as a boy.
>
> More importantly, I came to appreciate his character and beliefs—a patriotic, unassuming, dependable American—the kind that made our country great and will keep it that way despite national trends and fancies...[12]

I took a deep breath and looked up at the floor-to-ceiling bookcase beside me. The first thing I did in our new home was place the humidor holding Dad's ashes on a high shelf. On either side, I had carefully arranged all the books from his library about World War II and his Bible. Dick Taylor had retyped the war story we had discovered, and my mother had had it printed and bound. She distributed the blue booklet, entitled *Memories of World War II as Recounted by John Fletcher Sisson*, to friends and family.

Although I had yet to finish reading the booklet, it was of utmost importance to me that his story occupy a place of honor. A comfortable loveseat was placed adjacent to the bookcase and a faded wicker chair was placed in front of it, though no one was allowed to sit there.

Six months after my father died, Mom had been diagnosed with heart failure and was scheduled for open heart surgery. True to form, she faced it just like the two mastectomies she endured. She wore her stylish hospital gown, complained about the food, and told us all she loved us.

As my mother prepared for surgery, I turned 42 and was astonished to discover that despite a history of unsuccessful fertility treatments and miscarriages, I was pregnant. As if that wasn't enough, Gary announced he had accepted a contract to develop a radio station in Vero Beach, about two hundred miles south of my parents' home in Sawgrass.

By the first anniversary of my father's death, my mother had accepted our offer to join us in our new home in Vero Beach and sold the house in Sawgrass. I would spend the next five months reading books or taking walks with Gary and Mom on the beach as we all anticipated the baby's arrival. When a book captured my imagination, I often read late into the night on the sofa in our living room. Next to the sofa sat a red and green wicker accent chair that my mother's cat used as her favorite scratching post and place to sleep.

During those late, quiet hours, I came to the unsettling realization that my father also was sitting in the chair from time to time. At first, instead of being comforted by his presence I was completely unnerved. Gradually, I became protective and did not allow others to sit there. Finally, I began to nod or say "Dad?" when I sensed his presence. I assumed he was watching over Mom.

Life drifted along in this gentle way until I gave birth to Kathryn on October 1, 1992. And then, in short order, Gary completed his contract, and we decided to return north. This seemed the perfect opportunity to live in one of my favorite places, Andrew Wyeth's Chester County, Pennsylvania.

I don't remember when the cat became the sole occupant of the chair. But I do know this, one day my father started visiting and then one day he stopped. I imagine a psychologist could have a field day with that, but deep within me I knew something profound had taken place. I began sitting on the loveseat and occasionally talking with him as I looked at the chair or up toward the heavens and the humidor.

When the old chair was little more than a shredded shadow of its former self I reluctantly agreed to throw it away, but our conversations did not end.

Part Two

A QUESTION OF HONOR

"On 4th May 1945, the German surrender was signed at Montgomery's headquarters on Lüneburg Heath. This simple ceremony marked the end of the Third Reich, created by Adolf Hitler to last for a thousand years... although only twelve years had elapsed.

... those twelve years of preparation for and prosecution of a war for world domination plumbed the depths of human suffering."[13]

— **Barrie Pitt,** Introduction to *Himmler*, 1972

Heinrich Himmler visiting the Dachau Concentration Camp, Dachau, Germany, 1936[14]

5. A Deadly Legacy

WINTER 2004

"This book needs a home, Dad." I brandished Heinrich Himmler's personal copy of *Mein Kampf* toward the bookshelf as I tried to remember the old family story.

```
I was the officer in charge of securing
Heinrich Himmler's home at the end of
the war. There was a large, gold embossed
Mein Kampf on a pedestal next to his
desk. It was too large for my pack, so I
took the smaller volume from his desk.
When I got back to headquarters, I real-
ized it could go home with my commanding
officer or me. I decided to keep it.
```

As I considered the worn volume in my hand, I wondered why we had never asked him any questions about the story: Where was the house? Why was he there? I glanced up toward the humidor that still held his remains, his story skimmed but not read, and his books on World War II. He had died thirteen years ago this very month. In an odd way, I was relying on him even more since his passing, particularly when it came to helping my mother, who still lived with us.

"What do you think?"

I had retained the habit of discussing life with my father as I sat on the love seat beside the bookcase. Although he wasn't there, his quiet authority remained. I found courage in the integrity he valued in his life and his calm, ironic, reaction to duress. Lately, discussion centered on Gary's and my divorce. Dad heard a litany of late-night rants and tears as I confronted ending a 20-year marriage and my fears about my ability to resume a professional life.

Somehow, I found the gumption to return to school in 1999 at the age of 49 to pursue a master's degree in historic preservation. I had just begun working as the executive director of a small nonprofit in Wilmington, Delaware. Mom provided a loving constant in Kathryn's life as we both tried to help her through her own grief over her parents' separation.

The idea of dealing with Dad's old war booty gave me something I could focus my energy on, something to take my mind off all the things I had so little control over. I was further galvanized by what I read later that night in my father's copy of *Himmler* about the atrocities committed by the man who had owned that *Mein Kampf* and the evolution of the "largest instrument for mass terror the world has ever known."[15]

The next morning Mom looked surprised by my idea. "I never gave it much thought. It was your father's."

"The question is, how does one deal with Nazi memorabilia?"

Neither of us had a clue. We did not read German, but we could see that Himmler had signed his name on the front endpaper in an unsettling artistic script. My mother suggested I call

Sotheby's in New York. We knew nothing about authenticating books, particularly first editions owned by Nazi war criminals; the famed auction house would have an answer.

"Excuse me?" The voice did not hide its owner's consternation.

"I am trying to find out what the market is for a *Mein Kampf* owned by Heinrich Himmler."

"Excuse me?"

I repeated my request.

"Sotheby's does not deal in Nazi artifacts."

"Could you help guide me to the correct organization that does?"

"Sorry." We were disconnected.

It never had occurred to anyone in the family to take my father's war booty any more seriously than he appeared to. It was surprising that after all my parents' moves the book was still in their possession. It always seemed to find its way into a forgotten corner of a library shelf until, after they retired to Florida, it was relegated to my father's files. I opened the volume. What did the penciled notations in it mean?

After several more frustrating dead ends, I had an idea. Tucked into a curve on a back country road, announced by an unobtrusive sign, is the mammoth three-story structure that is "Baldwin's Book Barn." Entering, I could feel the great timbers sagging from the weight of the thousands of used books that rested inside, each reflecting the resiliency of the human spirit.

I set my book down on the counter, "This is Heinrich Himmler's copy of *Mein Kampf* that my father brought back from World War II. My family is trying to figure out what our options are for finding a new home for it."

The man took a step back from the book and me.

"Let me get the owner." I thought he was rather brusque.

"Thanks." I turned to browse the display of recently acquired books.

As a collector of first edition Oz books, I had purchased several volumes from the store. My father had introduced me to the series, returning from business trips with much anticipated new sequels. I collected the first editions because of this happy memory.

"Tom tells me you have a book you want me to look at." The owner was already holding the *Mein Kampf* and regarding it with curiosity.

"Yes. My father brought this back from World War II and my family is trying to figure out what to do with it."

He considered me for what felt like a long time. "And how did this happen? Do you mind telling me?"

"My dad passed away in 1991. I can only tell you what we were told. He was an infantry officer and was responsible for securing Heinrich Himmler's home at the end of the war. He said there was a large, gold-embossed *Mein Kampf* on a pedestal next to Himmler's desk, but it was too large for his pack, so he took the small volume on the desk."

He continued to look at the pages in silence. "Do you know what the handwritten notations are?"

"No. I don't read German and you are the first person who has examined the book outside of the family. I came here because I have purchased Oz books from you for my collection. Sotheby's said they did not deal in Nazi artifacts, and I thought you might be able to help us."

He finally stepped up to the desk and gave me a puzzled smile.

"Do you mind if I take the book to my office to see what I can find out by doing some research?

"Not at all."

After about twenty minutes, he returned with an article from the *Washington Post* dated December 8, 1998. It featured Richard Breitman, a history professor at American University in Washington who had written a biography of Himmler.

> Heinrich Himmler was just 27 when he studied a copy of Adolf Hitler's *Mein Kampf.*
>
> The man who would eventually order construction of the Auschwitz concentration camp evidently liked what he read.
>
> Himmler's copy of the *Mein Kampf* has been donated to the Museum of Jewish Heritage, and historians say his annotations reveal the ideology that propelled him to the head of the Nazi police.
>
> "It shows that the doctrine of antisemitism, nationalism and even how to get rid of the Jews was already being considered by Himmler as well as Hitler in the 1920s." Manhattan District Attorney Robert Morgenthau, chairman of the museum's board, said Friday.
>
> "He took the racial ideology lock, stock and barrel," Breitman said. "It's another line of continuity in early Nazi ideology to the policies of the Nazi regime and eventually to the Holocaust."
>
> The historian said scholars had been unaware of the book's existence.[16]

I looked with astonishment and then dismay at the article and then at the book which he had also returned to me. So, this couldn't be what my father said it was, Heinrich Himmler's personal copy of *Mein Kampf.* But he signed it!

The owner read my thoughts, "I did some research. Adolf Hitler first published *Mein Kampf* in two volumes. If your father's story is true, this is clearly Volume I because Volume II of the set has been donated to the Museum of Jewish Heritage. If this is the case, it could have a value of at least $20,000 on the legitimate book market. I don't know anything about the black market for Nazi artifacts."

For the first time in my life, I suddenly felt the magnitude of the book in my hand. I wanted to drop it. "My family would never profit off the blood of those who perished in World War II." A flush of moral superiority masked my guilt over not beginning this journey long ago.

He politely nodded. "I suggest you contact the museum." He gave me a second copy of the article as I tried to express how grateful I was for his help.

I returned to my car confused. How could all of us have ignored this grotesque reminder of the Holocaust? My father had clearly considered the book's symbolic meaning when he bought the companion book about Himmler and the Nazi atrocities he organized. But why leave the matter there, just sitting in a forgotten corner of our house?

Why not talk about history when you played a role in it. Why not share that history with your family?

I couldn't just pin this on my father. I also had coexisted with this book and the enormity of the crimes it represented and inspired. Heinrich Himmler's fearsome words whispered in my ear,

It is a crime against our own blood to ... assume they are human as we are human.[17]

6. With All Due Respect

SUMMER 2004

When I returned home, I told Mom all I had learned. She encouraged me to contact the museum, which I did a few days later, leaving a message for one of their curators. When I received a call back, I was prepared for the hesitancy in the man's voice. I explained the article in the newspaper and that we had Volume I of Himmler's *Mein Kampf*. We made an appointment for me to bring the book to New York the following week. Mom and I couldn't believe our progress, and since it involved an important museum, we decided to include the next generation in the project.

My daughter, Kathryn, and I arrived at our destination on a pleasant August afternoon. The Museum of Jewish Heritage — A Living Memorial to the Holocaust is located at the southernmost tip of Manhattan. It is an inspiring home for "New York's contribution to the global responsibility to never forget." The museum is committed to the "crucial mission of educating diverse visitors about Jewish life before, during, and after the Holocaust."[18]

A senior curator met us at the entrance. Walking through the solemn reminders of World War II Nazi atrocities I was struck by the fact that the visitor was also encouraged to look through the windows to the Statue of Liberty and Ellis Island. The curator was careful to point out important aspects of the exhibits to us. It felt very special to share this experience with my daughter.

We stopped in front of a small book. Beside it was a dedication which read, "Anonymous donation in special honor of 'the girl in the red coat.'" He turned to us and said, "This is Volume II of the *Mein Kampf* owned by Himmler which the article you mentioned referred to." I nodded, wondering what was behind the dedication.

Then Kathryn and I walked through one of those doors you see in a museum but are never allowed to open. It was just as I had imagined, an energetic, bustling, and cluttered world of museum administration. The curator took us to a small round table. Once seated, he asked if I would tell him my father's story.

When I finished, he asked, "May I see the book?"

I took it from a manilla envelope and handed it to him. He opened it, looked at several pages, then took a deep breath. He suddenly began to handle it like a delicate and fragile object.

"May I take it to compare it to the other volume?"

I nodded.

He finally returned with several documents. He looked at me with wonder. "I compared it to the other volume. There appears to be an exact match with Himmler's signature and the notations." I wanted to say, "Well what did you expect," but was respectful. I realized that the world of books and artifacts viewed me as a strange woman with a stranger story carrying an old red book around and asking for assistance. They didn't know me, and they certainly didn't know my father.

As we sat together, he began to explain the next steps he hoped my family would consider. The book, now carefully bound in a protective cover, rested by his side the entire time we talked. He told me I should contact the Himmler scholar Dr. Richard

Breitman to authenticate the book. He also told me he had contacted the family who made the initial anonymous donation and the donor wished to speak with me. I agreed he could share my telephone number.

In the next few moments, I experienced the love that curators put into their work as they preserve the artifacts of history that add so much to our understanding of our heritage. I could sense the anxiety he felt when he gave the book, now found, back to its owner. I suspected he was worried the United States Holocaust Memorial Museum in Washington, D.C. might be very interested in it. Having seen Volume II, my mind was made up about where Volume I belonged, but the decision was not mine to make.

When Kathryn and I returned home, Mom listened with interest to what had been suggested. She indicated she would talk with my brother and sister about donating the book. We talked at length about the importance of reuniting the two volumes. Kathryn shared how upsetting it was to see the children who were victims of the Holocaust. I put the book back in the curio cabinet and waited to hear from the anonymous donor.

"Mrs. Marshall?" The voice sounded crisp and pleasant.

"Yes."

"My name is Charles Williams (a pseudonym) and I understand from the museum that you may have the companion *Mein Kampf* to the one my family donated to it."

"Yes, that's correct. Thank you so much for calling me. My family was really at a loss as to what to do until we found the article about the donation of the second volume. To be honest, we had no idea that this was one of a pair."

"I would be happy to share all that I know, but I wondered if you would mind telling me how your family came to possess the book."

I told my story once again, this time with any embellishments I could remember in terms of Dad's military service.

There was a long, odd pause when I finished. And then he said in a careful and measured voice, "Mrs. Marshall, with all due respect, I believe your father lied to you."

Lied to me? I sat stunned, phone in hand. All my friendly exuberance evaporated. "Excuse me?" I resisted the urge to hang up.

The caller was undeterred, "Both Volumes I and II of Himmler's *Mein Kampf* were in my father's possession at the end of the war. Was your father assigned to intelligence? Could he have somehow taken Volume I from him? I didn't mean to upset you, but I have been in shock since the museum called."

"What?" Now I sounded like the lady at Sotheby's.

Silence.

I calmed my emotions, but my voice was still sharp. "My father was a lieutenant in the infantry. I don't know what you are talking about. He was a man of honor and a man of his word. He was also a man who picked up souvenirs all his life."

My father's propensity for souvenirs, even as a respected corporate lawyer, was a topic of great amusement for our family. I grew up thinking everyone ate breakfast with Danish modern stainless steel with airline logos on them. His most famous escapade was convincing the top executives at Columbia Gas to lift salt and pepper shakers from Trader Vic's in New York City. It never crossed my mind to question what he said.

"I just don't understand," the voice continued. I began to relax. I could tell he was as upset as I was. "I must admit I was confused when Volume I was not found with my father's papers after he died. I assumed he gave it to a close friend he served with during the war. When that friend recently died and no book appeared, I was baffled. The timing of your call to the museum is a remarkable coincidence."

His honesty disarmed me. "I am sure there is some kind of confusion, but I feel certain that my father's story is true. Are you sure your father told you both volumes were in his possession?"

There was another pause, "His stories could be vague, but that is what I always understood. As I told you, I remember that he said he had both volumes at the end of the war."

I took the opportunity to change the subject and asked if he would help my family make the donation. He readily agreed.

After I hung up the telephone, I walked to the love seat and sat in silent shock. What had I gotten myself and my family into? My father's integrity was one of the few constants in my life. And now this?!

Did someone fail to tell the truth? Well, it wasn't my father—of that I was certain—wasn't I? It was impossible to imagine that I, like Dorothy in the beloved books my father gave me, had found an imposter behind the curtain.

The cancer metastasized; he didn't tell you?

7. The Final Solution

FALL 2004

A month or so later, my second conversation with Mr. Williams went much better than our first. I was pleased to tell him I had an appointment in November to meet with Dr. Breitman to authenticate the book. Charles shared the story of how he had guided his family to make the donation of Volume II. His father had known Mr. Morgenthau, and his family was honored to contribute to furthering the museum's mission. He did not, however, address the question that I had wondered about since I first visited the museum. Who was the girl in the red coat? Instead of asking, I decided to wait. Before hanging up, we agreed to stay in touch.

I was aware of the haunting image of the little girl in a red coat who had symbolized the brutal and senseless death of millions of Jews in *Schindler's List,* Steven Spielberg's powerful film set in Nazi-occupied Poland. The girl's cruel murder in the movie seemed reflected in the dedication made by the Williams family, but I sensed the two were not related. Even so, I liked to think of the young girl's memory triumphing in the end over the dark shadow cast by the owner of the *Mein Kampf* and architect of the "Final Solution."

That evening, I climbed on the love seat and took the small

blue booklet that held my father's story off the shelf. The old ache of missing him returned. I reread his opening line.

```
This is Saturday night, the sixth of Jan-
uary 1945.
```

I was not going to find a quick answer to Mr. William's questions in my father's narrative. It stopped months before the war ended, which was when Dad said he took the *Mein Kampf.*

There were things I questioned about my father, but his integrity had never been one of them. He was my moral compass. My trust had been confirmed by the many notes the family received after his death. I had quoted one at his service in Darien:

> John was an able and exacting lawyer's lawyer. I admired and learned from him. He was a patrician and a vestige of what made our country and our shared profession great heritages. It was my distinct privilege to have had the opportunity to share a fraction of his life space.[19]

Two fathers, two volumes of the *Mein Kampf,* two conflicting stories; how to resolve the questions now that both men were dead? At least Dr. Breitman would be able to confirm my father's statement that the book had belonged to Himmler. I closed my mind to further speculation and returned the narrative to its shelf.

I met Dr. Breitman at a coffee shop in Baltimore's Fells Point, across the harbor from my old home in Federal Hill. He was an affable man who immediately made me feel at ease. We chatted about Baltimore and how much I had enjoyed living there. I recounted my father's story and handed him the book.

As I watched him quietly turn the pages, pausing now and then, it struck me how impossible it is to fully assess an individual. He had brown hair and glasses, and was slender, and comfortably dressed in a sporty manner. I tried to intuit the intelligence and discipline required to lead a team declassifying more than eight million pages of US government records in compliance with the Nazi War Crimes Disclosure Act. All this while teaching at American University, writing scholarly books, and serving as editor-in-chief of *Holocaust and Genocide Studies* in association with the United States Holocaust Memorial Museum.

He was pleased to tell me he felt the book was indeed Volume I of Heinrich Himmler's copy of *Mein Kampf*. He translated some of the notations for me and said that the conversation was between Himmler and his father. As with Volume II, the elder Himmler supported Hitler's philosophy but not the more radical solutions.

We agreed he would further review the book, return it by special delivery and contact the museum with his findings. I asked him about the large gold-embossed *Mein Kampf* my father had mentioned sitting on a stand beside Himmler's desk. He paused, then told me he had heard that such volumes existed but never had seen one. I felt encouraged by his response; it seemed to corroborate my father's story.

Less than two weeks later, I received a large package from American University. I opened it and found Dr. Breitman had returned the *Mein Kampf* and included a signed a copy of his recent book on Himmler, *The Architect of Genocide, Himmler and the Final Solution*. With the book also was the Spring 1999 issue of *Holocaust and Genocide Studies*, which contained his appraisal of Volume II, "*Mein Kampf* and the Himmler Family: Two Generations React to Hitler's Ideas."

While I waited for the formal assessment of Volume I, I decided to read Dr. Breitman's book. Not only did I have little understanding of Hitler's manifesto, I possessed limited knowledge about the specifics of the Holocaust and Heinrich Himmler's role in it. The book meticulously traced the evolution of the nightmare that unfolded as Himmler, under Hitler's leadership, developed a system to annihilate Europe's Jews and "impure races." A quote from the introduction of the book would reverberate in my conscience as I read. How could this have happened, I kept asking myself.

The commandant of Auschwitz once said:

> "Our system is so terrible that no one in the world will believe it to be possible...If someone should succeed in escaping from Auschwitz and in telling the world, the world will brand him as a fantastic liar..." That was nearly true during part of the war, and there is a danger that it may become true again.[20]

The first Nazi concentration camp, Dachau, opened in 1933 shortly after Adolf Hitler became Chancellor of Germany. At the end of the war, there were twenty-three main concentration camps. In all, Nazi officials are thought to have established over 40,000 incarceration centers for enemies of the state, forced labor, prisoners of war and death camps.

Breitman writes that Himmler and the SS had begun planning the extermination of Europe's Jewish population by 1939, and Hitler was fully participating by the summer of 1941. After a four-hour meeting with Himmler on November 15, 1941, one of Hitler's top aides, Alfred Rosenberg, held a confidential briefing and told German journalists that "the Final Solution had begun; it was a 'biological extermination of all Jews in Europe'...The press was not to write about the extermination in detail, but the reports could use stock phrases such as the *definite solution* or the *total solution* of the Jewish question."[21]

Two weeks later, on November 29, German police in Riga, Latvia segregated men between the ages of eighteen and sixty who were capable of labor and settled them in a part of the Riga ghetto, which was sealed off. That evening German policemen and Latvian auxiliary police raided the rest of the ghetto and assembled all the remaining Jews, killing or injuring many in the process. The next morning the same police used whips to drive the fifteen thousand assembled Jews out of the ghetto on what would be a death march. A trainload of a thousand Jews from Berlin had also arrived in Riga, the first of the transports from Germany. Some had frozen to death en route in the unheated railway cars. The survivors were forced to join the march of the Riga Jews to a nearby forest called Rumbula. An eyewitness described the slaughter that took place.[22]

> The Jews had to dig long trenches as mass graves, then undress themselves completely, sort out their clothes taken off in separate heaps, and then lie down naked at the bottom of the mass grave. Then they were shot by SS men with pistols. The next group of those condemned to death had to lie down on top of those already executed and were shot in the same manner. This procedure continued until the grave was filled. It was then covered with soil, and a steam roller was driven over it in order to flatten it out.[23]

A week later, at least another eight thousand people were killed the same way. When the success of the killing operation was reported to Himmler,

> ...there had already been complaints from military and civil authorities in the Ostland; the killings were not secret enough. So, Himmler said that shooting was too complicated an operation; it was better to

use the gas trucks...With that, the extermination camps became the vehicle to deliver the *final solution of the Jewish problem*.[24]

Although I wanted to stop reading many times, I finished the book. It was terrifying to confront what humanity can do and has done throughout our history. There was no longer any doubt in my mind about the importance of our family's pending donation to the Museum of Jewish Heritage.

Since my father, unbeknownst to him, had only taken Volume I of Himmler's copy of *Mein Kampf*, I was interested to learn more about both volumes. Our copy was published in July 1925, and it took Himmler until 1927 to finish it. The first volume was largely based on Hitler's personal experience, and Breitman said Himmler wrote that some parts were weak.

Breitman noted that Volume II, published in 1926, held the most practical value for Himmler because it focused on Hitler's ideology and the Nazi movement in general. He stated that the notations made in the second volume make it clear that as deputy leader of the SS, Himmler was looking for practical ways to apply the "truths" of *Mein Kampf* to the work of the SS.[25]

These practical applications remain almost incomprehensible. Breitman observes in his book's introduction, "This combination of sophisticated technology and barbaric mass murder raises serious questions whether there has really been progress in history, and it is a stark commentary on the human capacity for evil."[26]

Eyewitness and famed CBS correspondent Edward R. Murrow arrived at the Buchenwald concentration camp on April 12, 1944, the day after the 6th Armored Division of Patton's Third Army liberated it. During his broadcast about the camp, he seemed to understand instinctively how easy it is to reject a truth that is almost impossible to absorb, "I pray you to believe what

I have said about Buchenwald. I have reported what I saw and heard but only part of it. For most of it I have no words."[27]

Dr. Breitman notified me in early December that he had contacted the museum to confirm the authenticity of Volume I. At least that part of my father's story was put to rest. I prayed our family's gift could help the museum meet its vital mission of education and remembrance. For as Breitman had written,

> Those who are prejudiced or naïve may not try to absorb the events of the Holocaust at all; it seems to them common sense to be skeptical about the accounts of gas chambers and crematoria and millions of murdered Jews.[28]

I was left to ponder if I truly knew how the book came into our possession.

8. A Moment in Time

2005 – 2006

On November 15, 2005, a year and a half after I walked into Baldwin's Book Barn, Mr. Williams and I finalized our plans. My mother would present the book my father brought home from the war to Robert M. Morgenthau in his capacity as chairman of the board of the Museum of Jewish Heritage. We all agreed Mr. Williams should be at the meeting with my family.

Mr. Morgenthau was a legend in his time. His obituary in the New York Times in 2019 hailed him as "the courtly Knickerbocker patrician who waged war on crime for more than four decades as United States attorney, then Gotham's aristocratic Mr. District Attorney."[29] After New York City's prominent district attorney cleared a morning from his schedule with less than three weeks' notice, a date was set for December 2nd.

Just as we were preparing for the meeting, the whole adventure was derailed by a criminal crisis in New York City. The District Attorney's office was exceedingly apologetic. We contacted the museum and mother was advised to mail the letter she had planned to hand to Mr. Morgenthau. The actual donation of the book would have to wait. We all tried to mask our disappointment.

Her letter, dated December 2, 2005, reflected her wishes:

Dear Mr. Morgenthau:

I wish this letter to explain my gift to the Museum of Jewish Heritage of what is to my best understanding Volume I of the personal Mein Kampf owned by Heinrich Himmler. The book and handwritten notes made by Himmler and his father have been authenticated by Dr. Richard Breitman. I ask that any further academic or other questions about the book be referred to Dr. Breitman or other experts.

I make this gift with great pride and loving memory of my husband who was a lieutenant with the 22nd Infantry Regiment landing at Utah Beach on D-Day. He wrote a journal about that experience which I am also sharing with your staff. I believe that he would agree with my children and me that this book was always the property of the Jewish people and that it is our honor to return it not only to its rightful owners but to rejoin Volume II as a lesson that good can ultimately triumph over evil.

I ask that this gift be dedicated to an American infantry officer and his contribution to the ultimate triumph of conscience over Nazism in World War II. Your Director of Collections and Exhibitions has suggested a shortened dedication that reads: "In honor of a US infantry officer who helped conscience triumph over Nazism." This is very acceptable.

I wish you the very best in your work to convey the meaning of the Jewish experience and the horrors of the Holocaust to the world.

Sincerely,
Betty L. Sisson [30]

We finally rescheduled the donation for June 15, 2006, but we lost my mother in the process. She made it clear she was not going to try again to make the trip, despite all my pleas. She was beginning to feel her 84 years and insisted it had been my project from the start. Kathryn, my sister Elizabeth, and I reluctantly arranged to travel to New York without her. I would complete the donation to the museum while Kathryn went shopping with my old college roommate.

I was on time to meet Elizabeth at One Hogan Place, the old Criminal Courts Building on the edge of Chinatown. We were escorted to Mr. Morgenthau's office on the top floor, and he met us at the door. He was charming and gracious and was interested in the narrative my father had written. I later learned that a near-death experience as a young naval officer in World War II had led him to devote his life to endeavoring to "being useful."

Dad, are you here? Are you sitting with us at the great mahogany table in the center of the room? I hope so.

We exchanged stories until the secretary came in to say Mr. Williams sent his regrets. Acutely disappointed, I wondered if he decided not to confuse our moment of pride with the shadow of our father's crossed stories. Nonplussed as a career lawyer must learn to be (my father always lit his pipe at moments like this to gain a moment to think), New York's longest serving DA, mob buster, public avenger and feared prosecutor smiled at me as if we were old friends.

"I understand your family wishes to make a donation to the Museum of Jewish Heritage and that Elizabeth and you are here on behalf of your mother who is making the donation."

"That's correct." My hand shook a bit as I handed him the *Mein Kampf* and a copy of my father's World War II narrative.

The donation complete, Mr. Morgenthau escorted us to his

favorite Chinese restaurant, where he regaled Elizabeth and me with stories of New York. I was intrigued by the security detail that accompanied us. After we took our leave. Elizabeth and I high fived before she headed home to Connecticut. I went off to find my daughter, who had purchased a new pair of trendy flip flops. Over ice cream cones, my report on the day's events was well received.

Mr. Williams called that evening. He sounded tired. "I couldn't bring myself to join you. I hope you will forgive me. My family and I are so grateful for what you have done. I have been going through old paperwork to see if I can find some reference to resolve the mystery of how your father ended up with the book."

I told him how proud our family was to play a role in reuniting the volumes and recounted the details of the presentation and lunch with Mr. Morgenthau. We agreed to stay in touch. I knew we would not.

After everyone else was in bed, I went to sit on the love seat. I knew why Mr. Williams stayed home. He felt certain his father told him he had a book from the war in his possession that had been sitting on my family's bookshelf until today. We were both raised by men who valued personal integrity. Why would someone lie about this in the first place? It made no sense, and now both soldiers were gone. They could not defend themselves or reassure their children.

Suddenly my gut lurched. Is that why Dad never finished the journal he wrote in January 1945? I could not bear the thought of discovering that my father's silence was due to the fact he was fabricating part of his service record.

Dad, why aren't you here? What is the answer to Mr. William's question? You held us all to the same high standards of conduct you established for yourself. Or at least I thought you did.

For the first time since his death, I removed the humidor that contained his ashes from the special shelf. Nestling it in my lap, I cried. I was learning that contrary to many of the books and movies portraying it, very little about war offered a satisfying conclusion.

Two soldiers, spring 1945, two crossed stories.

I sat on the love seat in the dark silence for a long time.

Part Three

WADING INTO HISTORY

"War is three things. It is a way of deciding great differences among people. That is the politicians' war. It is a game of infinite complexity. That is the generals' war. And it is a personal experience of weariness, hatred, terror, filth, and occasional heroism and humor. That is the soldiers' war."[31]
— **Charles Christian Wertenbaker,** *Invasion,* 1944

A female demonstrator offers a flower to military police on guard at the Pentagon during an anti–Vietnam demonstration." S. Sgt. Albert R. Simpson, Arlington, Virginia, October 21, 1967 [32]

9 . Whispers of War

SPRING 2016

Five months after the *Mein Kampf* donation in 2006, my sister and I were attacked by the same enemy that had disfigured my mother and killed my father. We were diagnosed with cancer within months of each other. With my own battle to wage, the difficult question of my father's honor faded into the busy reality of life.

My sister and I survived, but if I sat on the love seat to talk to Dad during those years, it was to ask for prayers and courage, not guidance. Over the next decade, life offered little time to think or reflect. In addition to cancer, I faced divorce, established a new career, helped my mother through her last years of life, remarried and saw Kathryn successfully off to college.

It wasn't until 2016, three years after Mom's death at ninety-one, that I returned to the love seat for quiet meditation. Although my parents' ashes were now buried together, I found comfort as I gazed up at my father's books, still carefully arranged on his shelf. I was grateful to find this refuge still welcoming me.

I had decided it was time to organize my parents' belongings. One of my projects involved placing their personal papers and effects in two separate piles for careful consideration. Each had a collection of family photographs from the late 1800s and early 1900s.

With the photographs on the coffee table in front of me, I was trying to connect faces with names. I would hold a picture up now and then and have a conversation with the parent I thought they belonged to. My husband thought I had lost my mind but then decided it was amusing as it saved him from listening to the stories.

My father's things were easier to deal with because he, not my mother, had kept childhood scrapbooks filled with memorabilia from high school and college. He also had journals from his trips with his Aunt Kathryn and letters home prior to and during World War II. Dad's side of the family were writers, and there were several books and memoirs in my library. One letter from his grandfather to his wife caused me to pause. It ended, "My ambitions seem to center around my namesake John Fletcher, I hope he will never lose his self-poise or sturdy character." I put it in a special place.

My mother's pile contained a few pictures from her teens—largely with her first cousins or several boyfriends—and many pictures of pets and horseback riding. There were one or two modeling pictures. Mom had little chance to form fond recollections of her school years; she was shy, and her family moved close to twenty times during her father's military career. There were, however, her pastel and watercolor paintings, her poems, and family photo albums that were bulging with pictures and memorabilia. I always thought she kept them for herself, but now I wondered if Dad had inspired her.

Dad's scrapbooks revealed a young man who was at the top of his game in high school and college. There was no hint of his disappointment in not attending Dartmouth College after his father lost his fortune in the Great Depression and the family faced economic hardship as he worked to rebuild his lumber business. My father settled happily into Ohio State University where he played sports, was active in a large fraternity, and headed up schoolwide events. This included the all-important Homecoming Queen festivities. Dad and his friends all had leadership positions. When they returned from the war, I could tell from the scrapbooks that the party started up again right where it had left off.

I surmised that my mother, a New York model, must have been an intriguing change for my father. Although she retained several good friends and had many interesting stories, she made it clear that she had sought old-fashioned standards that New York City did not offer. I wondered if this handsome, active, young man with his wide circle of old friends offered her romance and the stability her life had denied her. I know their union was grounded in true love. The few letters I read from the Korean War years, right after they married, were proof enough. The scrapbook my father started in the fall of 1947 after they first met also showed he knew Bette was special. No more homecoming queens.

But there was something else I discovered among those memories. The wars nations wage wove like a faded ribbon through their lives. In addition to my father's World War II narrative, there were numerous reminders of the impact of war on our family that I had previously ignored. A shadow box held medals from two world wars that belonged to my maternal grandfather, Col. Earl Lyons, and my father. My mother had kept her father's dress military jacket and framed his letter from General Eisenhower congratulating him on his retirement from active military service. There were pictures of Dad's Uncle Horace Sisson, a captain in World War I. He tragically died shortly after returning home from the effects of being gassed. I found a picture of Horace's brother Paul also in military uniform.

My father had kept two military rifles, a German dagger and Luger, and his duffle bag from World War II. They were in the two green wooden boxes used to ship his belongings home after the Korean War. The gifts from Japan that he brought home with him were displayed in our house. I found several folders of correspondence and newsletters from the 22nd Infantry Regimental Society and two World War II scrapbooks, which I added to the library I had created. A scrapbook and yearbook from the Korean War showed my father, was now a major with the 37th Infantry Regiment headquarters.

Near the very end of my organizing, I found a poem Mom wrote to me for my twenty-second birthday. I was still in a muddle, between communes and college, working to find a balance between the traditional life I grew up in and the counterculture I had embraced. Dad may have managed the financial and ethical underpinnings for our family, but Mom graced us with imagination and an abiding love of nature and the arts. The poem ends:

> *Then college came*
> *And what a jump*
> *Sometimes my throat*
> *Had quite a lump*
>
> *Sit downs—strikes*
> *Demonstrations and radicals*
> *Protest marches*
> *Indecision and sabbaticals.*
>
> *With a disposition only*
> *God can give*
> *A rebel with an ache*
> *To live!*

And tho I've had
Some sleepless nights
Some due to croup
Some adventurous flights.

I'll still say those years
Have been the best
And in you
I am truly blessed.

Memories of the late 1960s came flooding back. My war years, so radically different from my family's. After seeing my historical connection to the Army, I realized how personal my arguments must have been to my parents. My father, grandfathers, and uncles risked or lost their lives in service to their country. My father and mother had never mentioned that.

10. Two Worlds Collide

1968 – 1969

It was 1968, the deadliest year of the Vietnam War for American soldiers and the beginning of the loss of widespread American support for the war effort. I was aware of the anti-war protests and the 1967 "Summer of Love" in San Francisco's Haight-Ashbury, but I was a firm believer in the domino theory and supported the war and my conservative family during my high school years in Darien, Connecticut.

That fall I arrived at Skidmore College (for women) in a green Villager wool skirt, a flowered McMullen blouse, stockings and matching Pappagallo shoes purchased at Bergdorf Goodman. When I came home for Thanksgiving, I was braless, wearing Navy-issued denim bell bottoms, a Carnaby Street shirt and knee-high boots. At a party with old high school friends, an upperclassman who knew me well poured his beer over my head. He called me a "communist, pinko pig." Two of the boys my own age rose to my defense and drove me home. I showed them how to smoke hash.

By Christmas, I had read Tom Wolfe's *Electric Kool-Aid Acid Test* and Norman Mailer's *Why Are We In Vietnam?* and my dominos tumbled. I went full-on anti-establishment, anti-war, pro-love. On Saturday, January 18, 1969, I visited a friend from home at Williams College in Williamstown, Massachusetts. One of his friends told us there was going to be a Counter Inaugural Ball in Washington, D.C. to protest Richard Nixon's inauguration as President of the United States. The National Mobilization Committee to End the War in Vietnam (MOBE), a broad-based anti-war coalition, was organizing the event to decry the pro-war platform of the new president. Did we want to join those attending from Williams on Sunday?

And so, without a change of clothes or proper walking shoes, I was swept up in my first demonstration against the Vietnam War. Street theater, kazoos, peace signs and smiles were the order of the day. "Make love, not war" seemed to be our guiding principle. The signs and banners reflected the broad base of protesters: *Abolish the Draft, Free Political Prisoners, End the War, Bring the Boys Home, Don't Eat Grapes, Defeat Imperialism Everywhere,* and *Victory to the Vietcong.*

Amid the organized groups that all had signs and often looked and dressed alike, Johnny and I fell into the category of lost preppies. I was such a neophyte that I wondered why we were cheering for the National Football League, "Ho Ho Ho Chi Minh, NFL is gonna win." Halfway through the march this girlfriend of a high school football star realized we were shouting NLF—National Liberation Front.

There was one tense moment when supporters of the President confronted the marchers, but otherwise we ended up at the Counter Inaugural Ball in a large tent at the foot of the Washington Monument without incident. I heard that a pig by the name of Pigasus (the titular leader of the "Yippies") was to be In-Hogurated, but if that happened I never witnessed it. I do remember the cold and the mud. You could see the stage and light show but couldn't really hear the speakers and performers. We talked and smoked pot with the people around us. The confusion was overwhelming but exhilarating; we were taking a stand for freedom.

On Monday the protestors planned to challenge Nixon's motorcade, but we had to head home. Johnny's friends took us as far as Baltimore, then we hitchhiked back to Darien. We got as far as the Darien service station on the Connecticut Turnpike thanks to a generous family who scrunched together to make room for us. My feet hurt so much I silently blessed them and did my best to entertain the ten-year-old boy who sat with us in the back seat. He thought we were beyond cool.

Long after midnight, we called our parents for rides home. My father listened silently as I described all that I had seen, making no mention of the late hour and the classes I would miss the next day. In the morning, he told me to take my VW Beetle back to Skidmore. When I worried that freshmen were not allowed to have cars on campus, he told me to park it somewhere away from the school.

At Skidmore, I learned the demonstrations on Monday were considered a success. Police were outraged over the burning of American flags and arrested several demonstrators. But there was no repeat of the previous summer's violent clashes between protestors and police during the Democratic Convention in Chicago.

After a year of more peace rallies and late nights smoking legal and illegal substances while debating life's meaning, what began as an adventure turned into a chasm of doubt. The endless questions about the war and why it was being waged upset all I had once held to be true. In the winter of 1970, I dropped out of college, told my parents to sell my clothes and headed west in a used white Plymouth Valiant Dad bought for me to find my answers. An old friend and co-pilot painted a huge clown's grinning face on the trunk of the car.

Although the song would not be written for another two years, Graham Nash's battle cry for the Chicago Seven perfectly summarized my philosophy, "We can change the World...It's dying, rules and regulations, who needs them..."[33]

In three months, I went from living in a free church and panhandling on the streets of Boulder, Colorado to hitchhiking up Highway 1 in California with my change of clothes in a sack and a black cat on my shoulder. "High on life" was my battle cry, and I went in search of my Holy Grail too broke to do drugs or buy gas and miraculously protected from sexual predators. After a series of adventures, which included raising Atlantis with Ken Kesey and his reunited Merry Pranksters, I returned to Boulder only to be arrested for vagrancy while camping out in a communal cave in the foothills of the Rockies. After a night in jail with a witch and a shoplifter, I agreed to a $50 fine and was released. When I called my parents for help, my father said, 'I will pay the fine, but you have to figure out how to get home." I somehow managed.

The Vietnam War and Richard Nixon's presidency eventually came to an end. The cultural world of the 1950s I had been born into disappeared into the tailwinds kicked up by their demise. I would spend the next decade, after I finally graduated from college, trying to find a way to integrate my youthful idealism with the realities of adult life. My father and I had warm and loving moments, but then would clash noisily. Questions surrounding war remained a source of contention.

Years later, my mother said Dad felt our generation was denied its childhood by the cultural upheaval over the war. I don't like to think about what they felt as they watched me drive away in that Valiant bound for nowhere. During the years Mom and

I would spend raising my daughter Kathryn together, I did have time to make amends. Not so with Dad. As close as we were, there was always an uneasy truce.

Now, sitting on the love seat, I wondered if the uneasy truce was only mine? After I got home from the West Coast, Dad kept my car, huge grinning clown and all. He drove it to work after they moved to Wilmington so my sister could use his yellow VW for school. Driving the car clearly amused him. He recounted how his secretary had told him in no uncertain terms that that jalopy was not appropriate for a man of his stature.

You just don't understand!!

Was it only me who didn't understand?

11. The Soldiers' War

FALL 2016

The rush of memories from organizing my parents' papers revived the uncertainties surrounding my father's acquisition of Heinrich Himmler's copy of Volume 1 of *Mein Kampf*. My husband, Hunt Bartine, and I were busy building a life together, but that summer Mr. William's unnerving accusation agitated me like a small wound that just wouldn't heal. I asked myself, what right does anyone have to accuse a man who is dead of a crime that he cannot defend himself against?

With all due respect Mrs. Marshall, I believe your father lied to you.

The more I considered what little I knew about my father's service in the war, the more I began to feel like I had dropped into a courtroom drama where two reliable witnesses had conflicting versions of what happened on "the night of the crime." There was, after all, little reason to question the veracity of either man. Yet if their children had remembered their words truly, someone was lying on the stand. But why?

I came to recognize I had little interest in questioning Mr. Williams and his father. All I really wanted to do was prove my father's innocence in the matter; to resolve the issue of honor and integrity which I had always associated so closely with him. Was there a mystery to be solved? I began to wonder if it was possible to prove my father's story to be true.

Sitting on the love seat and gazing up toward Dad's books, I realized I had not once taken the time to finish reading his account of his experiences during World War II. On hearing about his manuscript, many of my friends would say, "I wish my dad had written something down. No matter how often I asked, he would seldom talk about the war." They only knew little bits and pieces.

Why, Dad, why didn't you talk to us about what you really experienced in the war—or at least why you took the Mein Kampf and what you intended to do with it? Or maybe tell us why you returned to the Chateau? How can we honor your memory when we don't really know what you did? Or prove you told the truth... or understand why you didn't?

Could his wartime account, even though it ended in January 1945 answer some of my questions?

Standing on the love seat, I took the small, blue-covered copy of my father's narrative off the shelf. It was so frustrating to be unable to relate to many of the situations he described, and his military jargon always confused me. I had journeyed to Portsmouth, New Hampshire to research the story of my mother's family homestead. As the Preservation Officer for Chester County, I was meticulously documenting the footsteps of the American and Crown Force armies during the Battle of Brandywine in 1777. How could I ignore my father's contribution to history?

For some reason, my mind drifted back to a day in 1994, the 50th anniversary of D-Day, three years after my father died. I was working at the front desk of a YMCA, wearing one of his two Purple Hearts to honor his service and memory. I proudly explained how my father had earned the medal after landing at Utah Beach on D-Day to all who asked. I was speaking to a man about my age when a shocked look crossed his face, and he seized my collar and started to pull me over the counter.

"What are you doing!" he shouted. His friend interceded and it was over in a second. I dropped back stunned. "What are you doing?!" he loudly repeated.

Still surprised I mumbled, "What am I doing?"

"That's a Purple Heart." His voice was broken and jagged. People began to gather around us. His friend looked sternly at me, "No one wears a soldier's Purple Heart but the soldier." He then turned to speak with his friend.

I hastily removed the medal from my shirt and said something about being very sorry, that I didn't know, and that I was trying to honor my father. I carefully placed it in my purse and left the desk. I felt confused and ashamed, and I was not about to continue the conversation.

"He probably served in Vietnam, and it left him weird." I could feel the old anger returning. We had been right to march against the war and raise our fists. How many petitions had I signed saying I would burn my draft card if I were a man?

Eldridge Cleaver, the writer and political activist, said in 1968, "There is no more neutrality in the world. You either have to be part of the solution, or you're going to be part of the problem." I chose to be part of what I considered to be the only solution. Anyone associated with the war (along with essentially anyone over thirty, as Jack Weinberg of the Free Speech Movement at Berkeley famously told a reporter in 1964) was part of the problem. That included my father.

Thinking back on the exchange with the vet, I decided to investigate the meaning behind the Purple Heart. U.S. service members of any rank who had been wounded or killed in combat were eligible. Gallantry in battle, honorable service, heroic actions, and solemn distinction — all words I had never associated with war were reflected in the medal first awarded by George Washington during the Revolution.

What were my opinions on war and soldiers? Did I still believe that all war was bad and soldiers at worst were part of the problem and at best innocent dupes? If that was true, where did this leave Dad and his military service. And me, a veteran's daughter.

How does one learn about being a soldier if no one who was one talks about it?

I looked up at the shelf that once safeguarded my father's ashes. I knew it was impossible, but I was hoping for an answer. He did speak to me once after all. I thought about the day after my father's death in 1991 when I had discovered the file cabinet filled with memories of World War II and the Korean War. I again felt my confused surprise at finding he had lived an active military life, one that his family knew little about. Dad certainly had had no trouble telling Bill Weingart, his old squad sergeant, about what he did and remembered in the letter he wrote about his trip to Normandy in 1979.

Could it be that soldiers only talk to the men they fought with?

Idly paging through the small blue volume, I wondered if I could figure out how to understand my father's experiences during World War II and perhaps become better acquainted with the man who never shared that part of his life with his family. Could that in turn inform me about decisions he made — specifically concerning a book he "picked up" toward the end of the war? Could I prove that he had not dishonored the Purple Hearts he deserved?

To unravel the "mystery" of Dad's *Mein Kampf* and the lingering question of his honor, I decided I needed first to understand my father better as a soldier. With this decision, the historian in me knew what to do. I would undertake a study, just as I would for any historic event or property.

I did not plan to become an expert on World War II, or even D-Day for that matter, but I needed to have a working understanding of Operation Overlord and my father's place in the historic landing at Utah Beach on D-Day. The next step was to carefully read the war experiences he recorded within that context. I also decided that the best way to do this was to edit his narrative for clarity, so that I, a civilian, could better understand his words.

After considerable debate with myself, I decided I would only use books relating to his experiences that my father had clearly read. That evening, I climbed onto the love seat and removed each of the volumes related to World War II from his shelf and placed them beside his narrative. They were dusty. I hadn't touched one of them since our move in 1993.

The first selection was easy. *The History of the Twenty-Second United States Infantry in World War II*, compiled and edited by Dr. William S. Boice, Chaplain. It had a faded blue paper cover and was dated by my father, July 28, 1959.

I established a second criterion with *Eisenhower, At War, 1943–1945* by David Eisenhower. My father had made several notes in the book. I decided to include only volumes that showed his interaction with them. I rejected Omar N. Bradley's *A Soldier's Story* because it was not annotated. I also didn't know who Bradley was.

The Longest Day by Cornelius Ryan[34], which inspired the movie of the same name, was stamped "John F. Sisson." I decided to read this book first based on Mr. Ryan's statement, "This is not a military history. It is the story of people, the men of the Allied forces, the enemy they fought and the civilians who were caught up in the bloody confusion of D-Day." I noticed Dad indicated the men he knew who had been interviewed for the book. Chaplain Boice was one of them.

On the inside cover of *D-Day Beaches Revisited*, by Patrice Boussel, my father had written: John F. Sisson, 2nd Lt., 2nd Battalion, F Co., 22nd Inf., Utah Beach, June 6, 1944. I paused to run my finger over the words and felt the melancholy that reminded me he was gone. The author was French; the French edition was published in 1964, the English version in 1966.

Invasion! by Charles Christian Wertenbaker had been published in the fall of 1944 "This is the first connected story of the most stupendous military and naval operation in history, the Allied invasion of France." Wertenbaker, reporter Ernie Pyle and *Life* photographer Robert Capa had been assigned to General Bradley's headquarters and "rode with our advance troops into the villages of Normandy while German bullets spattered death in the narrow streets."[35] There were not only annotations by my father but six faded newspaper articles from the same period. Wertenbaker was my third eyewitness, joining Chaplain Boice and my father. I excitedly put the book on top of my pile.

I sheepishly returned to *A Soldier's Story* and added it to my collection, annotated or not. Now that I realized Bradley had commanded the American forces on D-Day I needed to include him. It was ludicrously clear how little I knew about the war my dad fought in—or for that matter, the military in general. The remainder of my father's collection of World War II books had been gifts from his children, acknowledgments of the military service we knew so little about. I put them back on the shelf.

It did not take me long to realize that reading military history without a working military vocabulary is essentially a waste of time. Armies, divisions, regiments, battalions, and companies, with a dizzying array of numbers and letters attached to them, moved around through those histories with great speed. I had no idea how it all related. Every other word seemed to be a military abbreviation, jargon or slang. No wonder I never could finish my father's narrative. My confusion only grew worse the more I read. In desperation, I returned to the inscription where Dad told me exactly where he stood within the great invading Allied force. One man, one uniform, "John F. Sisson, 2nd Lt., 2nd Battalion, F Co., 22nd Inf., Utah Beach, June 6, 1944." What was an Inf., a Co., and a Battalion?

After many starts and stops, I developed a series of charts, glossaries, and maps to help me better understand the context of my father's story. These included an outline of the army command my father served under during World War II and a timeline of action by the 4th Division's 22nd Infantry Regiment in 1944 (my dad's story did not extend into 1945). Creating a guide to the army lingo in Dad's manuscript was very helpful as were maps I found depicting troop movements. (I have included a glossary of military terms and slang as they appear in his story to help other civilians negotiate the terminology.)

Because my father's narrative focused primarily on the build up to and execution of D-Day, I decided I also needed an overview of that battle. I turned to three sources that were not in my father's collection. I already owned *D-Day June 6, 1944: The Climactic Battle of World War II*, by Stephen Ambrose, and *Utah Beach*, by Joseph Balkoski. Both had been published shortly after my father's death. The third was a series of books on World War II published by the Center of Military History, the United States Army. *Utah Beach to Cherbourg* contained many of the maps I later found in my father's files, but that's getting ahead of my story.

12. A Story Unfolds

FEBRUARY 2018

By Christmas, I had completed my background reading and a slightly revised version of my father's manuscript that at last allowed me to fully appreciate his story. I felt like I had climbed a mountain backwards. I had finally started to bridge the gap between my father's version of World War II for his family, which as I said sounded more like the experiences of Gregory Peck in *Roman Holiday*, and the following article in the *Columbus Dispatch* announcing why he was home on leave from Europe with the time to write about his recent experiences.

Jan. 30, 1945

Among the 1,300 decorated veterans who arrived home recently from the western front in Europe was First Lieut. John F. Sisson, who is spending part of his 30-day leave with his parents, Mr. and Mrs. Warren B. Sisson, 315 Gill Avenue (Galion, Ohio)…The returned veterans were pulled out of the battle line, under fire in Germany, France, Belgium, and Holland over a month ago to return home for thirty days. Every one of them had been wounded once or twice—some four times or wore two or more decorations for valor…Lieut. Sisson came here from Camp Atterbury, Ind…He wears the Purple Heart with an Oak Leaf Cluster, two stars on his European area ribbon and the Combat Infantryman's Badge.[36]

I intended to print my update of his manuscript and distribute it to family members for the 27th anniversary of his death with the following introduction. Instead, my father's narrative became the foundation for this book.

* * *

D-Day, June 6, 1944, was the beginning of the end of World War II in Western Europe. The worldwide hostilities had begun in September 1931 when Japan invaded Manchuria and ended in 1945 with the surrender of Germany on May 8 and of Japan on September 2. It is estimated that over 3 percent of the world's population died during the war.

Germany, Italy, and Japan were the core expansionist Axis powers, although other countries joined and left the alliance during the war. In Europe, Adolf Hitler and the National Socialists, or Nazis, came to power in Germany in 1933, and soon began the persecution of Jews, political opponents, trade unionists, homosexuals, Roma, and many other groups. In 1936, Germany signed an alliance with Italy, ruled by the Fascist dictator Benito Mussolini, to form the Rome-Berlin Axis, and then signed an anti-Soviet pact with Japan, which Italy joined in 1937. Despite the latter agreement, Germany signed a nonaggression treaty with the Soviet Union in 1939; a secret clause in this treaty called for Germany and the Soviet Union to divide Poland between them.

World War II began in Europe on September 1, 1939, when Germany invaded Poland. Honoring their guarantee of Poland's borders, England and France declared war on Germany two days later. Germany, Italy, and Japan signed the Tripartite Pact in 1940, and in June 1941, Germany invaded the Soviet Union. On December 7 and 8, 1941, Japan attacked the United States and the British Empire in the Pacific and Southeast Asia, and the United States declared war on Japan. After Germany and Italy then declared war on the United States, Americans found themselves fighting a war across two hemispheres, with the Axis powers controlling large portions of Europe, North Africa, and East Asia.

When the United States entered the war, Germany occupied Czechoslovakia, Poland, Denmark, Norway, Holland, Luxembourg, Belgium, and France in Europe. During the summer of 1941, Germany had attacked Russia and the Axis armies in North Africa were threatening to drive through Egypt to the oil lands in the Middle East.

Winston Churchill and Franklin Roosevelt had had secret discussions before the United States entered the war, and a formal alliance of Britain, America, the Soviet Union, and 23 other nations—the United Nations, more commonly called the Allies—was signed on January 1, 1942. On May 11, 1942, Allied forces landed in W North Africa, where British forces in the east had been fighting the Italians and then the Germans since mid 1940. Victory there led to the invasion of Italy, which began in Sicily on July 9, 1943; landings in southern Italy took place on September 3. Allied leaders committed to a landing at Normandy during strategy talks in December 1943. They initially set May 1, 1944 as the date of the invasion. By the late winter of that year, more than two million troops from over 12 countries were in Britain in preparation. The invasion of northwest Europe was code-named Operation Overlord.

U.S. Gen. Dwight David Eisenhower was named Supreme

Allied Commander, and British Gen. Bernard Law Montgomery commanded all Allied Ground Troops. Lt. Gen. Omar Nelson Bradley commanded the U.S. First Army, whose VII Corps (consisting of the 4th Infantry Division and others) would land at Utah Beach and whose V Corps (consisting of the 1st and 29th Infantry Divisions) would land at Omaha Beach. Lt. Gen. Miles Dempsey commanded the British Second Army, whose 30th Corps (consisting of the 50th Infantry Division) would land at Gold and whose 1st Corps would land at Juno (the 3rd Canadian Infantry Division) and Sword (the British 3rd Infantry Division). The U.S. 82nd and 101st Airborne Divisions were to be dropped inland of Utah Beach and the British 6th Airborne Division inland and east of Sword Beach, near Caen.

It is generally agreed that the invasion of Normandy on June 6, 1944, was the largest amphibious invasion in the history of warfare. More than 132,000 Allied infantry soldiers landed at the five beachheads on Normandy's English Channel coast, and an additional 23,400 airborne troops landed inland using parachutes and gliders. Severe weather postponed the attack by a day, and weather conditions hampered some of the airdrops. The attack began shortly after midnight when the airborne troops were dropped to secure key inland objectives and hold them until the infantry reached them. Nearly 7,000 naval vessels ferried troops and supplies or bombarded the beaches while some 11,500 Allied aircraft supported the landings, dropping airborne troops and striking enemy ground targets.

On D-Day, 23,000 soldiers landed at Utah Beach starting at daybreak under the command of the United States 4th Infantry Division. They suffered 197 casualties — killed in action. At Omaha Beach the United States 1st and 29th Infantry Divisions, totaling 34,000 soldiers, suffered 2,400 casualties. At Sword

Beach, the British 2nd Army, 1st Corps, with 29,000 soldiers, suffered 630 casualties; at Juno Beach, the 3rd Canadian Infantry Division and British 1st Corps, totaling 21,400, experienced 1,200 casualties; and at Gold Beach, the British 2nd Army, 30th Corps, with 25,000 soldiers, had 400 casualties.[37]

Casualties were high among the airborne forces who landed in the dark during the early morning hours of D-Day. They were essentially surrounded by the enemy until they linked up with the amphibious forces. The U.S. 82nd and 101st Airborne Divisions suffered 2,499 casualties at Utah Beach, where their successes were largely responsible for the low casualties suffered by the amphibious forces, and the British 6th Airborne suffered 1,166 casualties at Gold Beach. A reporter with NBC turned to look at General Eisenhower as the 101st Airborne lifted off into the air that night and "saw tears in the Supreme Commander's eyes."[38]

Utah Beach was the westernmost of the five beachheads, located at the base of the Cotentin Peninsula. The 4th Infantry Division's objective was to secure Cherbourg, at the northern end of the peninsula. Its deep-water port could be used to supply Allied forces. David Roderick, who also landed at Utah Beach, wrote a memoir to "give credit to all the sailors, airmen and airborne soldiers who helped make the landing of my division such a success."[39] In the end,

> D-Day was a smashing success for the 4th Infantry Division and its attached units. Nearly all objectives were attained, even though the plan had to be abandoned before the first assault waves hit the beach. It went on to fight battles far more costly than the one it won on the Cotentin beach on June 6, distinguishing itself throughout the campaign in northwest Europe, especially in taking Cherbourg, in

holding the German counteroffensive at Mortain, in the liberation of Paris, in the Hürtgen Forest, and in the Battle of the Bulge.[40]

The three infantry regiments that commanded the rifle companies of the 4th Infantry Division were the 8th, 12th and 22nd and each numbered close to 3,000 men. Each infantry regiment was in turn divided into three battalions of 1,000 men, the First, Second and Third. Each battalion had three rifle companies and a weapons company, First Battalion: A, B, C, D; Second Battalion: E, F, G, H; and Third Battalion: I, K, L, M. Each company consisted of 190–240 men. Of the soldiers in each battalion, 875 were riflemen; the balance were headquarters, medical and service detachments. Each rifle company was divided into 41-man platoons composed of three 12-man squads plus their platoon leader (usually a lieutenant) and his team. A sergeant commanded each squad. This was the offensive power of each infantry regiment.[41]

The 4th Division's 8th Infantry Regiment's Second Battalion's E Company was the first Allied company to land at Utah Beach, starting at 6:30 A.M. Brig. Gen. Theodore Roosevelt, Jr., was in the first boat, and would be attributed with the famous statement, "We'll start the war from right here," after it became clear they were over a mile south of their original objective. For his actions that day, he earned the first Medal of Honor for the Division.

My father's 22nd Infantry Regiment began landing when its Third Battalion came ashore with the 8th Infantry Regiment beginning at 7:45 A.M. The First and Second Battalions followed, after 9:30 A.M. The platoon of forty-one men my dad commanded led F Company which, in turn, led the Second Battalion ashore around 10:00 am. Their mission was to attack towards the northwest to reduce the strong points at Crisbecq and Azeville with the First Battalion.

The troops waded inland through lowlands the Germans had flooded. Men plunged into water over their heads if they stepped wrong. They reached dry land in the vicinity of Saint-Martin-de-Varreville and Saint-Germain-de-Varreville. Encountering more resistance than expected, they did not reach their objectives by nightfall.

Stephen Ambrose wrote, "Totalitarian fanaticism and discipline would always conquer democratic liberalism and softness. Of that Hitler was sure." He then eloquently credited the soldiers who fought at D-Day, and I would add each day after until the final German surrender on May 8, 1945:

> But for all that American industrial brawn and organizational ability could do, for all that the British and Canadians and other allies could contribute, for all the plans and preparations, for all the brilliance of the deception scheme, for all the inspired leadership, in the end success or failure in Operation Overlord came down to a relatively small number of junior officers, noncoms, and privates or seamen in the American, British, and Canadian armies, navies, air forces, and coast guards...
>
> It all came down to a bunch of eighteen-to-twenty-eight-year-olds. They were magnificently trained and equipped and supported, but only a few of them had ever been in combat...They were citizen-soldiers, not professionals.
>
> ...But when the test came, when freedom had to be fought for or abandoned, they fought. They were the soldiers of democracy...and to them we owe our freedom.[42]

During January 1945, one soldier (my father) recorded his role and the role of his infantry regiment in contributing to history, then never shared it. Until now.

Part Four

ONE SOLDIER'S STORY, 1943-1944

"We could see Omaha beach off to our left—our own Utah beach dead ahead. Battleships were sending up great flashes of smoke and flame as they threw their shells shoreward. Rocket boats disappeared in a cloud of smoke as they discharged their deck full of rocket launchers…

Suddenly there were no more questions. Only a G.I. saying, "those b------- are trying to kill us sure as hell." That is all that is needed to make a soldier fight. That is what makes him do brave things or cowardly things. Just that one phrase.[43]

— **2nd Lt. John Sisson,** D-Day, June 6, 1944

13. Welcome to the 4th Infantry Division

1943–1944

John Fletcher Sisson, my father, hailed from Columbus and Galion, Ohio. Born in 1917, he turned twenty-two two months after Germany invaded Poland. He graduated from Upper Arlington High School in 1935, Ohio State University in political science in 1939, and University of Michigan law school in 1943.

As war clouds engulfed Europe, the United States 4th Infantry Division was reactivated on June 1, 1940, at Fort Benning, Georgia. The move in September 1943 to Camp Gordon Johnston, Florida gave the division realistic amphibious training in preparation for the assault on Fortress Europe.

Dad began ROTC training during law school and entered active service on March 13, 1943. He was trained as an infantry unit commander at Camp Robinson, Arkansas, Camp Fannin, Texas, Fort Benning, Georgia and Camp Gordon Johnson, Florida. Reading his letters home, I learned his most pressing concern was buying a car and then receiving gas ration cards from his parents. He had this to say about his introduction to the army in his personal journal:

```
A beginning always seems like an end,
while you're living it. That's about how
```

March 1943 was. An end of school customs and thinking. An end to careless hours, libraries, flannel trousers, drinking beer the civilian way. But there were new customs and thoughts. New careless hours—not so many libraries or flannel trousers, but plenty of beer. Basically, and technically, I didn't know enough to be a good Pfc. When I hit the army, I resented my first captain like hell because he knew I didn't have much knowledge of the Army. That's the beginning. Now I know he was right. I would like to tell him so should I ever see him.

"What did you do in school?" the captain demanded immediately, referring to military training.

Of course, I had probably learned more law than he would ever see in his life and had developed a helluva lot of theories on beauty and comfort and living that were as good as they were different than his. But I was just finishing up being a schoolboy and wasn't enough sold on myself yet. My main reaction was to gape and let my knees knock—knowing full well I hadn't done anything he was interested in. He probably thought I might eventually be able to write my name, but that it was doubtful I could do much better than one inch block letters for a year.

```
Beginning number one. Be sure of your-
self.
How I longed for a command!⁴⁴
```

Dad was 27 when he landed in Normandy, France with the First Army, 4th Infantry Division, 22nd Infantry Regiment, Second Battalion, F Company on D-Day. He was a second lieutenant, platoon leader for the Third Platoon. He was 28, a first lieutenant, home on leave from active service in Europe with two Purple Hearts when he recorded his experiences.

I pictured him at the desk in his room in Galion in January 1945, after all the excitement had died down from his surprise visit home. I imagined that the young man who would become my father was thinking about F Company and the 22nd Infantry, who were still fighting hard through Europe while he composed and wrote the following narrative. I don't think he would mind that I am sharing it with friends and strangers, years after his death. I am pretty sure he would tell me I was going to a lot of trouble for nothing much. I would disagree, as I usually did.

Chapter breaks have been added along with my thoughts as I worked on each section. Other than that, this is my father's story as written.

```
This is Saturday night, the 6th of January 1945.
    Just about a year ago this story, or whatev-
er it can be called, started. Since every sec-
ond letter from Aunt Kathryn tells me I ought
to keep a record of this period, I'll say it's
to satisfy her—but really it is a pretty dull
Saturday night, and if Sunday morning turns out
better, this will undoubtedly never be finished.
```

Sometime in November 1943, the Green Dream—1941 convertible Ford model—and I pulled into a little known and better forgotten spot along the Gulf of Mexico called Carrabelle, Florida, home to Camp Gordon Johnston, U.S. Army Amphibious Training Center. I had picked up a couple of AWOLs from the 8th Infantry Regiment, the 4th Division being my first outfit. I was so busy casually trying to extract a little dope as to what I might expect that I didn't notice the numerous small craft along the shore and anchored a short distance out.

After the AWOL's (Absent Without Leave)] climbed out I noticed a few of these, and before I had completed the two-mile drive along the shore to Division Headquarters, I had begun to notice that all these craft had one common characteristic. The front of each was built so that it would drop down the square bow thus forming a ramp over which the contents of the boat might move. When I happened to notice one of these ramps drop down to disgorge a pack of running doughboys, I got the picture. From that time on it wasn't so much a question of whether we would be in the then, coming invasion of the continent—but more a question of which wave I could come in on.

A month at Carrabelle, taken up mainly with amphibious training, did nothing to alter this picture. We were learning the intricacies of assault teams, loading tables, flame throwers,

LCVP's, LCM's, Special Engineer Brigades, etc., all of which made up the amphibious training. Looking at it as a whole, it was a wonder of minute planning and clockwork timing. Problems were difficult enough, clearly the training wasn't just to pass the time.

November 25th, I believe, was Thanksgiving of 1943. On the night before, the 4th Division, in toto, was loaded into various sorts of small landing craft and taken out into the Gulf in groups of waves, as each landing section was called. The CT (Combat Team) 22 plus attachments such as artillery, engineers, reconnaissance units, etc.—hit the beach at H-hour, and H-hour hit daybreak. Having seen the plan for the problem, I was partially prepared for our watery reception—but not entirely.

The part of the order I had seen was the G-2 (Intelligence) report. This report described the beaches along the whole of the Carrabelle area. Only one section was described as full of sand bars offshore as far as 1,500 yards with 4 to 6 feet of water over the area. This section alone was listed as definitely unsuitable for amphibious landing by troops from small boats. This section was singled out as the worst landing beach for twenty miles in either direction. This section, of course, was Uncle-Red Beach, on which CT 22 was ordered to land on D-Day, H-Hour.

A definite strain was put on leadership. One

lieutenant, making the most of the position and attempting to imbue his platoon with some spirit of the "real thing," after an all-night ride in an open boat some 40 feet long, coverless and in one of the worst gales the Gulf had seen in several years, held his carbine over his head, pointed toward the almost invisible shore in the distance and plunged over the ramp with that immortal cry, "Come on men—this is it." Somewhere about "is" or "it" his battle cry turned into a salty gurgle and the poor guy disappeared in his own splash.

Although the ramp was resting on the bottom, only 3 feet under water, he had stepped off into an unforeseen hole which was at least 6 feet deep. I saw him three hours later, waddling down the road on land like a saddle happy cowboy, trying to keep his assortment of long johns, undershorts and wool OD's from chafing his legs off. The spirit was definitely out of his leadership, and he was quietly cursing the guy who wrote in the book, in small print at the bottom of the page, that an officer always sets the example, regardless of the cost. The cost was getting terrific.

The true G.I. spirit came to light. One numb, dripping, shivering, equipment-laden soldier was disgustedly throwing his pack on the beach in order to pull off his life preserver, when an immaculate Navy officer with enough rank to make him aware of himself appeared and inquired briskly. "Soldier, what is the situation?" Undoubtedly, he expected the soldier to try to orient himself and give some sort of an answer as to where he was and where he was going. The Navy Blue & Gold must have been confused when the G.I., without hesitating in his surly efforts to shrug off the life preserver, replied bitterly, "Sir, the situation is tactical." The Blue & Gold at least had the sense to admit defeat and got the hell out of there. (In infantry parlance: Fire and fall back.)

The true chicken---- spirit came to light. The Battalion C.O. happened to be in an LCTX which sank somewhere enroute. No one was lost, but all the equipment went down. There was an M-7 aboard the craft—a great track vehicle which mounted a 105-mm howitzer piece. Also, the Colonel's bed roll. According to instructions, all equipment had been attached to one cork on the end of a long cord. This was to act as a marker in case the equipment was lost. The next day found the Field Artillery out in a boat looking for the cork which marked their M-7. They might have been successful if the Colonel hadn't prevented any semblance of order by coming aboard and shouting orders and directions as he looked in vain for the cork which marked his bedroll. Well, after all, he had paid $35 for his sleeping bag. Maybe the taxpayers did have to pay twenty-two thousand for the M-7, but there were plenty of taxpayers.

I don't remember much else except vague recollections of mysterious and sinuous movements which pulled a blanket a little closer around me during the night when the waves were breaking over the bow of our craft. Somehow, the slightest movement of the little finger became all important in cutting off the wind. When I gave up trying to keep warm a few hours later, things were much better and I suffered in silence, comforted by the slightest breath of warm air from the engine, or by a single drop of spray being caught by the blanket.

As we approached the shore at first light on Thanksgiving morning, I peered over the top of the ramp, and between dousings from the waves could see that it was still entirely too far from shore to walk comfortably when dry, when our boat rubbed the bottom and ground to a stop on the sand. The coxswain yelled down that he would back it up and get over this first sand bar. His encouraging voice made shore seem a lot closer. Sure enough, he did get us over the sand bar. My joy knew no bounds, and I could see us balling up General Barton's G-2 staff and my own dire predictions as to this landing, by making an unprecedented approach to the actual shore.

I now believe that thoughts can flash across one's mind in a second, for as my mind's eye was getting several hours ahead of us the boat's bottom was making contact for the second and final time. We had progressed all of fifteen feet over the first sand bar, and this time had made it as far as we could. I don't remember any six-foot holes of water, but I do remember my legs were numb from one end to the other. My teeth were chattering so hard when I hit shore some three quarters of a mile later that I found myself giving my first orders by Morse Code. I was commended by the Bat. C.O., for the excellent initiation of automatic weapons my platoon put up, however.

After that brisk walk before breakfast, even canned C rations were a delicacy for Thanksgiving dinner. I did hear a few more optimistic characters mutter under their breath, "Fly, damn you, fly!" as they munched away at their erstwhile turkey. They probably voted the Republican ticket, anyhow, and were only making a concession to custom. Few of them would have cared to have taken the trouble to distinguish between cold meat and beans (or hash, according to their luck) and the real turkey McCoy.

The sun was shining though. I definitely remember that.

Incidentally, I don't want you to think that I was too advanced as a soldier myself. I had a platoon, but it was for the first time. I sent out three two-man patrols to contact the adjacent regiment, which

unbeknown to us had floundered around in the storm and become generally so disorganized it couldn't land on schedule. Of these six, I got one back, and spent three hours after midnight searching for the others. The problem was then over.

They showed up the next morning, and I'll be damned if one of them hadn't taken command and refused to allow the others to come in because, "the lieutenant said to find the 12th Infantry before we reported back." The 12th was water-bound until three in the afternoon and water-clogged after that but how was this joe to know it. He had his orders. On such stuff are lieutenants "brought up."

Major Kenan was the regimental S-3 at that time, and he sent me out to look for them. The look in his eye when I told him I wasn't too long out of ROTC was neither disgusted nor approving nor sympathetic. When I reported back to the battalion just about twelve months later in Luxembourg, he greeted me as battalion commander. I never appreciated a friendly smile more than the one he gave me there. Army memories are sometimes too damn long to suit me, especially when the lesson has been learned. I remember I approved highly of now Colonel Kenan's memory.

Fort Jackson, South Carolina is more or less hazy. So many things happened so fast there that I never did get them straight. I was officially given the third platoon of F Company. I knew by now that we were preparing to go overseas, and that as soon as the equipping was through, we would be on our way. Mother and Dad came down for a few days just before Christmas, and we had a swell visit. As good as any I can remember. I had trouble keeping a couple of late dates, but I think I did Mother and Dad justice.

Nancy came down shortly after they left, so I didn't have any more trouble with the late dates. A Christmas tree with presents around it, and a bottle of champagne almost did it. I'm still single though. I remember telling her at the train in Charleston, after her plane had been grounded by bad weather, that she shouldn't get worried for another week. I knew then that we were going to a port of embarkation at New York, but like Renfrew of the Mounted, honor of the regiment and all that, I couldn't divulge military secrets.

Once we reached the P.O.E. our whereabouts were supposed to be even more secret. But I kept my faith. I wouldn't tell her where we were but told her to drive me out to New Brunswick in the dead of night. Camp Kilmer is exactly three-quarter miles from New Brunswick. It took me ten minutes to walk past the line of cars parked outside the gate where most of the 4th Division boys were saying goodnight to their wives, sweethearts, or reasonable facsimiles thereof.

The ship wasn't sunk coming over, so there mustn't have been a slip of the lip. I came closer to being sunk when Sergeant Conly and Sergeant Mead broke the ice at Kilmer one morning and started trading punches over my body. That was the first time I blew my top at a soldier, and I guess it was effective. At any rate there wasn't any further trouble along that line, although my platoon sergeant and platoon guide were never much impressed by the other. I was glad they got the point that we had to have a good team, even at the expense of individual feelings. We had one, too.

By the time we embarked on the H.M.S. Capetown Castle, on the 19th of January, I could quote the Table of Organization—T/O to the army—of an infantry company in my sleep; could inspect my platoon in ranks and make corrections in the same condition; could quote the items listed on QM Forms 32 and 33 (Clothing and Equipment records) by heart and knew most of the men's clothing sizes better than my own. For eight weeks we had done little but check clothing and equipment.

When we were not doing this, we were consolidating clothing and equipment reports or planning new ones. I was getting my introduction to army poker at about this time, too. Cookie, Mr. Keep, Mr. Tiner and Dobro (of _if_-my-grandmother-had-worn-pants-she-would-have-been-my-grandfather fame) were in on this…and contrary to all reports it was not expensive at all. I only made 25 bucks, as I remember. I made the mistake of bragging about it and haven't won at poker since.

This was also the time of the famous shavetail cry, "Officers and second lieutenants." At a regimental meeting one night, the executive officer announced that all officers could now leave—then as an afterthought said that all second lieutenants would stay for a few minutes. From then on it was "officers and second lieutenants."

Captain Fulton was just married and wouldn't take a pass, because he could think about his wife just as well in his quarters as wandering out around the bright lights of the big city. If Jerry Claing hadn't lived in Hartford and Dutch Schultz hadn't decided to go up to Syracuse, where he lived, or if I had discovered the hole in the fence a little sooner, I might have had more time off. As it was, I kept it legal although we did take off three hours early on pass one afternoon after having just received an extension of several hours from Colonel Williams, our Battalion C.O. When we ran into him at the train station again three hours early, we wound our watches for a while and suddenly realized he wasn't supposed to be there either. We boarded the train together in a new-found fraternal bond.

Happily, my cousin Ellis was in New York, overseas bound too. We must have sailed in the same convoy, although it was months before we got around to writing to each other in England. I doubt if he was bucking
 as hard for his captaincy as I was for my sil-

ver bar. He made his though—so we accomplished something by not wasting time on correspondence. They weren't kidding when they sang, "There'll be no promotions this side of the ocean, So buck up my lads, bless 'em all." They weren't talking about "this side of the channel" though.

The boat trip over was a long series of boat drills, duty shifts in the troop compartments and poker games. American bills are no good in England and it is a lot of trouble to change them, so I simplified the whole thing and lost all mine trying to draw a small pair against jacks or better openers in a table stakes game. Benchley said that his junior year in college taught him that it was impossible to fill an inside straight. I would like to recommend that the universal college poker course extend its curriculum to the matter of drawing to small pairs, also. If they can't teach you how to make money in college, they can surely tell you how to keep what you have. Since my boat trip, I have been known to draw to a pair of tens when I'm running a terrific hot streak and the limit is ten cents (six pence, five francs or one mark.)

"Rags" made his first appearance as master of ceremonies of the 22nd Infantry talent nights while aboard ship. He must have been around before, but I hadn't been, so it was a first for me. The last time I saw him was in a barn in Krinkelt, Germany still saying, "O.K. fellas—O.K."

while he presented "Turn, Sports!" or the "Sheik of Araby." This ought to be a tradition with the 22nd after this war is over. They've been doing the same acts every time there is a show, and the Joe's love it.

We were in a convoy, being shepherded by the battleship Texas. The sailors said she was too old for active duty and wasn't much good even for convoy duty. The next time I saw her she didn't look exactly feeble though. It was about 7 am on the morning of June 6, 1944. The Cherbourg Peninsula wasn't target practice and the Texas' batteries weren't getting much chance to cool. It was Utah Beach, but Texas was damned if another state was going to get all the credit.

14. Foreign Soil

WINTER 1943-1944

While working on this chapter, I couldn't help thinking that the war sent Dad on a second trip to England which wasn't true for most of the men on his ship. I smiled when, towards the chapter's end, I read his nod to the "grand tour" he had made in 1935 with his Aunt Kathryn and cousin Ellis Phillips. Not mentioned, but certainly clear in his memory, was the Ford Phaeton, a four-door touring car that Ellis and he shared with Katherine Hepburn's younger sisters, while the mothers were chauffeured in Mrs. Hepburn's car. A far cry from 1944. He also mentions Hildreth, Kathryn's sister. Married to actor Nicholas Joy, she led a sophisticated life. That connection came in handy on his second "visit" to London.

I tried to imagine the odd contrast Dad's adventures in 1935 made with his new duties as a second lieutenant and the challenges they presented. Dad loved irony. I could see it in the stories he chose to tell.

```
We hit Liverpool around February 1, 1944.
    From the ship, one could see the results of
the '40-'41 blitz and the subsequent repair
work. There were untold tons of every sort of
```

equipment being unloaded, numberless ships of all sorts, the old buildings of the city behind us and a considerable part of England's largest port. All the G.I.'s saw that interested them much, though, were some women with painted legs (decent stockings being something you read about in England), cars driving on the wrong side of the road and a land that wasn't America.

Sergeant Mead, my platoon sergeant, took one look from the deck, from which not much could be seen, and snorted, "Lieutenant, this is a helluva country." This had variations that never ceased. Such as, "They ought to cut the cables on the barrage balloons and let the island sink." "I know now, why Hitler didn't invade this country. Even the Germans didn't want it." Or when he was particularly bitter, "The damn English tried to talk Hitler into taking over this place back in 1940, but he was too smart for them—so now they're going over to the continent to fight him and make him come back and take it."

I think someone cut his throat with his girl in Newton Abbot though, because when I saw him at the rehabilitation hospital at Leamington in the summer, he was behind England to the end, and his girl in Leamington felt the same way about it. Of such stuff are international relations made. And I'm serious. It is amazing how many fellows who would never have the slightest desire or inclination to be educated into a respect for Britain, got the general idea that for all the differences, there are some very important and commendable likenesses, too. And nine times out of ten it was the same way. Isn't it Ladies Home Journal that says, "Never underestimate the power of the women?" They'd better patent it or the State Department will be grabbing it up for a slogan.

A blonde, Red Cross girl who handed out an endless supply of hot doughnuts and warming smiles made a bigger impression on the men than the history or sights of Liverpool. There was a Kappa there from Whitman College, too. These gals had what it took—all of them we ever saw.

Three of us platoon leaders piled into a third-class compartment on the train, Shulzy, Lommler and me. It was a swell ride, we each had one musette bag; one blanket roll; one gas mask; one steel helmet, complete; one carbine, one val-pac; one paper bag full of alleged lunch; and an extra pair of shoes. All crammed full. Multiplied by three, it just about took care of the compartment without us counted in. The men had it good. There were only six of them in a compartment.

The trip started in the morning and lasted until well past midnight. Being new to the ETO (European Theatre of Operations)—and I find it hard to understand that there are still a large number of people who don't understand the translation of ETO as it is really much more famous than Spain—we were torn between a desire to appear nonchalant, and an inner feeling that we ought to carry our carbines on the alert and helmets strapped on in readiness for the air attack we were bound to get. After all, we hadn't had time to buy the evening paper or a war almanac, so how were we to know that there had been no raids in this area for about a year.

We didn't get away from the ship incidentally, without one good brush with the MP's. All the men were tired from the trip and from waiting most of the morning to debark. When we hit the dock, we were in no mood to argue. I had told Sgt. Mead to fall the platoon in, in a column of threes. SOP stuff. When I got there, he was ready to fix bayonets and assault the position. The MP said, "Fall in in column of fours." Mead swore back that he took orders only from the lieutenant and he said column of threes.

The MP was trying not to let too many people hear where the lieutenant and Mead might go, when I happened to come up. I was tired myself, and Mead and I had an unwritten law that one never let the other down. I let go with what wind I had left and told the MP that I had indeed told the platoon to fall in in column of threes and that what the hell army did he think he was working with to fall in column of fours. I told him this was the American army, and even if he had been Anglicized and perhaps denaturalized, we still fell in in column of threes.

By damn, this tickled Mead so much, especially about the American army, that he went ahead and fell the platoon in in column of fours anyhow and the MP fell back in confusion.

Welcome to merrie England, where they fall in in column of fours; drive on the left; where they call wheat, corn, and corn, maize; where the P-51 is a British plane; and no one has an icebox because no one did before. There were plenty of differences, and it took plenty of time to straighten them out.

I still maintain the Ladies Home Journal has the solution. If you ever meet an ordinary G.I. who says he likes England, ten to one he knows a nice girl in some little town there, who gave him tea or supper after Joe found out how scarce these are, and who was a pretty good sort of a chum even though she had been blasted out of two or three homes by bombs, served a year in the WRENS, ATS, Land Army or some other branch of the forces, and was at the time living as an evacuee from London, without home or family. Not all their stories were that bad, but it wasn't uncommon.

We landed in Newton Abbot, Devonshire, in the black of the night. The convoy got lost on the way to camp. That was fine because Shulz, Lommler and I were crammed in the back of a jeep and were lost anyhow. We stumbled around Denbury Camp for half an hour and finally got inside where there were some lights. Blackout shades looked unfamiliar, but welcome. (We were still waiting for the air raid.)

At the mess hall, we found out that British sergeants no more eat with the soldiers than do officers. Our American sergeants were looking pleased and proud as Punch at a private table with a special waiter. They got out their mess kits and sweated out the chow line with the men the next day, though.

America had now come to some corner of a foreign field that might forever be England, but the field was due for a little plowing. As a matter of fact, there were quite a few fields plowed, and the crop that comes out of it in England after the war should be very interesting. I think the doughboys will have done a Luther Burbank in a small way, and some hybrid blossoms on the old English stalk will no doubt appear.

We slept in the next morning until ten o'clock! Unheard of! Then we had had it. Calisthenics at 0800 daily. Close order drill at 0830. Classes until dinner. A hike from 1300 to 1600 and retreat. The moors came and went. The sleeping bag I bought on my last day in New York was tested, re-tested and super-tested. It never let me down. Never did a man have a more faithful friend.

The wind blew so hard at Hay Tor near Haytor Vale village that you could hear the babble of voices from Brooklyn, blown straight across the Atlantic in cold storage, and served up to you without a break. The map said the elevation was 1500 feet above sea level, but water ran out of the ground—when it didn't freeze. I put an outpost out around the base of the hill and those guys were really blue when they reported in the morning. Glad I couldn't see myself. Got into my post graduate course in poker with Cpt. Fulton, Buck Camper, George Samaris, Jim Beam and some of the others. They sent me to London for a two-day pass once, but I'm sure I repaid it several times.

The London trip was marked mainly by an application of the English caste system. Aunt Hildreth and Nick told me to look up an old friend of theirs who produces shows in London, and who was known to lunch often at the Savoy. Unable to find a room, I hiked myself to the Savoy in search of same. They had none—but I inquired about the friend. The name rang a bell—the clerk hummphed and hawwwed a bit and suggested I return in half an hour to inquire further as to both the room and the friend. Having seen the light burn, I left my bag with the porter, tipped him a pack of cigarettes, and took off down whatever avenue or street it is to see how much 1944 London looked like 1935 London.

It was much the same, until you made a closer examination. Then you would see long, long lines of buildings with sound looking fronts, but entirely boarded up. Entrances or unboarded sections as often as not revealed that the entire interior had been burned out. Block after block of this. Ironically enough, one of the first such places I saw was Berkeley Square, where the nightingale is supposed to sing. The only nightingales were some U.S. nurses coming out of a renovated building just off the square that the Red Cross had taken over for their use. The American Officer's Red Cross was just up this street, too.

In deference to better days, I ate again at Simpson's in the Strand as I had with Aunt Kathryn. Didn't see Ye Olde Cheshire Cheese or the Women's Club or Oxley Road, but went to Liberty's again, and walked by the entrance to Buckingham Palace where the guard had stood in 1935 with his white horse, silver helmet, white plume and scarlet tunic. A Tommy was there now, drab battle dress, tin hat and Enfield rifle, loaded and locked. The gates looked the same but a little more tired. They told me that they still changed the guard at the palace, as ever, but I didn't see it. London definitely had taken a terrific beating. I would hate to see some of the German cities that are taking it now and will continue to take it and which will have no chance to clean it up.

I made it back to camp for Monday morning formation with ten minutes to spare.

I was beginning to picture my father as a soldier. The contents of the file drawer that I opened the day after his death, all those military papers that I found, were taking on new meaning for me. I was getting to know and like some of the men who formed the backbone of the 22nd Infantry Regiment's Second Battalion.

15. Training on the Heath and Shore

SPRING 1944

While the politicians and generals debated, the men assigned to do the fighting were trying to get down to business in the spring of 1944. As deadly serious as the training they participated in was, I couldn't help but smile as I read those parts of my father's narrative that seemed to confirm a common British expression of the time, "Only three problems with the Americans. They are overpaid, oversexed, and over here."[45]

The region of England my father describes was an ideal training ground for the U.S. invasion forces. The 4th Infantry Division's location at Newton Abbot provided access to the untamed Dartmoor moorlands to the west and the English Channel to the east. Slapton Sands was a resort community on the Channel that was evacuated in April 1944 so that the area could be used for practice exercises for the landing at Utah Beach.

```
We ran some more problems on the moors—mostly
artillery overhead fire.
    On one problem, out near Princeton, the 105's
and 155's were laying down about 200-300 yards
```

in front of us and the 57's firing direct fire over us, we were picking handkerchiefs out of our hip pockets. We were learning. Gen. Bradley came out for that one, supposedly only for an hour—but he stayed for three hours and commended Maj. Edwards on the battalion's performance. Maj. Edward's first problem as battalion commander too. Which tells me that I'm getting ahead of the story by several Slapton Sands problems.

To make our training as complete as possible, a huge area off the coast was taken as a training range for our troops. The fleets of landing craft were beginning to assemble by early spring and by late spring the harbors of the south coast were clogged with naval craft of every size and description. Combat Team 22, 2nd Battalion, was training on LCT's. This is a landing craft that looks somewhat like a small yacht, with a pointed bow and the outlines of a seaworthy boat. Somewhat over a hundred feet long, I believe, it will carry around 200 men below decks. A gangway was run out and lowered from either side of the bow, and landing was accomplished by this means.

Considerably better than the LCVP. Things looked a little better. They rolled like a tub, though, and on one problem at least 90% of the men were to call it home. Beaver problem, Otter problems, unnamed problems, Tiger exercise, etc.

One of the wildest trips was with the 8th Infantry Regiment, part of the 4th Division. We were assigned to umpire one of their problems. We were to land in LCVP's from an LCT offshore. The first time we went in was a dry run, no ammunition to be fired. It looked like the wrong beach when we came in, but the problem went ahead. The one piece of equipment which could be used was a flamethrower and it promptly went into action on a small hut that looked like one of the prefabricated pillboxes set up on the ranges. It burned beautifully, and the operator was begging his lieutenant for permission to try it out on a house just beyond. Everything was bought up by the government, so what's one house to the war.

At just about this time, the Company Commander decided he was definitely on the wrong beach. According to plan, he began to make his way toward the correct beach, following the sea as much as possible. Suddenly, an English major appeared from nowhere and the captain could be seen gesturing frantically toward the correct beach, which was in view, and then back to us. Finally he stopped, looked up at the cliffs above us and started back.

It turned out that we had landed under the guns of better than a company of special guards who had loaded guns of every caliber and orders to shoot if fired on, or if a certain area they were guarding was trespassed upon. If we had had our allotment of ammunition, instead of a dry run, they would have opened up on us, and we might have had a preview of how effective a landing could be made against a well dug in and prepared enemy. Zowie—what a battle I almost saw. Better than any I saw on D-Day, I would bet.

The next day we hit the correct beach, but from the way those joes fired I wasn't sure if it was any safer than the day before. Ricochets were flying all over the place, a burst of BAR bullets went so close to my ear that their cracks as they went through the air made my ears ring for hours after—the area was full of dud rifle grenades and bazooka shells. But the payoff came after dinner. All morning the reserve assault team had lugged their ammunition, weapons, Bangalore torpedoes, grenades, bazookas, etc., along, with not a chance to fire. The rest of the company had fired their allotment of ammunition and were travelling light now. This did not set well at all with the reserve team. What they wanted was equality of weight.

Just after dinner our Company Commander gave the reserve team the mission of advancing about a mile by fire and movement. Expecting them to move under continuous supporting fire, I went along with them. The tactics looked rather strange, because the lieutenant was bunching his men up on a small hilltop. Finally, they were laying elbow to elbow in a line and the lieutenant gave his order to "start the attack."

"O.K., boys," he said, "let'er go." I've heard some noise in my life, but all those rifles, BAR's, bazookas, rifle grenades, machine guns, and whatever else they could find to set off, made one of the best I've run across yet.

A bazooka man fired a round into the top of a tree seventy feet away. The top of the tree—a big oak tree—fell out. This was a definite success even though he was firing over the heads of his own men. Rifle grenade men pointed their grenades without looking and pulled the trigger. One grenade went directly through a tree but somehow didn't go off. They fired over the safety boundary and finally were firing into their own troops. Those guys were undoubtedly nuts. They just naturally didn't give a damn. They hit the beach at H-Hour on D-Day and took and held more than their share of the beach. Of such stuff are winning armies made.

These problems progressed from a simple sort of exercise in which the troops embarked on the landing craft from some port or "hard", as the stone or concrete beachings were called. After a ride to the beach, the troops were put ashore by waves, and then went through a dry run of the problem. Enemy and firing were simulated—but equipment was very real—and heavy.

On one of these dry runs, Sgt. Durham, who was almost as strong as the famous Bull, had carried a Bangalore torpedo for several hours. The march had covered several miles of very hilly terrain, and the sergeant was getting tired of simulating. On this problem the artillery was firing a few rounds at targets ahead of the infantry. Suddenly, just as F and E Companies cleared the top of a hill, there was a terrific explosion between them, and company commanders came running from all directions, swearing that the artillery was dropping rounds short on us. Things calmed down shortly, and as we trudged away, I noticed Sgt. Durham looking considerably happier, minus his load of torpedo. Probably the only time an infantry man has been accused of being a 105 since the German tank commander reportedly gave up his company of panzers in Africa when the bazooka was first used and the Jerry thought a battery of 105's was zeroed in on him.

Jim Beam was our company executive officer, and his good humor and ability got us over many a rough spot. I don't believe I ever saw him either grin or scowl at the wrong time—except when he cleaned me in a particularly juicy poker game. I never did figure out his approach to English girls. It was infallible.

Our problems at Slapton Sands were reaching the stage of full-dress rehearsal. We hit the beaches firing everything we had, after preparation by naval gun fire and our artillery M-7's from their landing craft. On one of the latter problems, all the rank in the ETO showed up, from Eisenhower on down. Whether they were after our battalion CO before we landed or not, we have never been able to decide. At any rate, shortly after F Company had landed and proceeded to the top of the first ridge where we dug in to hold, the colonel came raging up the hill. With much southern accent, waving of hands, and wiping of perspiration from the brow, he demanded to know why in the hell the men had followed the hedgerows from the beach instead of proceeding across open fields.

The situation was indeed tactical that day. Furthermore, he had just been standing at attention before Generals Bradley, Montgomery, Collins, and Barton, which pretty well took care of things from Army Group on down through Division commanders. The book says somewhere that troops must get off the road under fire, as it is a natural target.

I had my men along a small dirt road no more than twelve feet across and bounded on either side by three- or four-foot hedgerows which offered perfect cover. It was a road though, and in keeping with the tactical situation he got me on this point. There being no other place for the men to obtain cover, I deployed them across a field, which he declared was much better. It was two weeks later that Gen. Bradley commended Maj. Edwards, our new battalion CO for the performance at the artillery problem on the moors.

Tiger Exercise was coming up toward the end of April. This was to be our real dress rehearsal for the invasion, with all the stage props and a complete cast. About a week before the exercise was to go off, I was called up to regiment, and informed that I would report to 7th Corps in Paignton, for Special Duty during the exercise. Not only this, but Class A uniform was required, and a jeep was furnished. I was beginning to get leery. From an inborn suspicion, I couldn't decide whether it was a frame up that would lead from a week or so of luxury to an assignment to land alone in France two weeks before the first troops, or whether my name had been confused with some visiting colonel. Off we went though.

Lt. Hellman was also assigned on the duty. 7th Corps met us and greeted us in the form of Maj. Brewer, a West Pointer of '40 and as fine a guy as he was capable. He explained our duties on the way to our hotel and told us to come back the next day. After several months of daily schedules, this was a little hard to take. Furthermore, when it turned out that the hotel was the only army mess in southern England that served ice cream nightly, we knew that we were in.

Except for the fact that there was too much rank in the hotel, which causes many an army nurse's eye to wander—even after you've spent the afternoon giving them a personally conducted jeep tour of the surrounding countryside—things were lovely. Torquay, which was next to Paignton, was a very pleasant place. One of England's swank sea resorts, it had suffered somewhat from the war, but was naturally beautiful.

The Imperial Hotel frowned on the American invasion but was making so much money that it failed to post notices that we were unwelcome. They merely raised the prices a few hundred percent and ignored us when they were not insulting us. Not all of them, of course, just the old dyed-in-the-wool types. I guess they like their system, I saw the most expert bowing and scraping exhibited for the benefit of a huffy old gentleman who deigned to toss about five pennies to two or three bell boys and the hall porter who stood in humble respect outside his car in the rain after loading a few dozen bags into his car. Anything less than two shillings for a Yank was taken as practically an insult.

The Torquay Hotel was much chummier after we brought George, the hall porter, around to our way of thinking. The scotch began to flow, and rooms were let at legal prices. A no-limit poker game with coins clanging on a hard top table one Sunday afternoon in the main lounge was rather hard for them to swallow. The Redcliffe Hotel was a success after we broke the orchestra in on "Embraceable You."

Tiger Exercise, as I said, was to be the full-dress rehearsal for the invasion. Naturally, everyone who was not directly concerned with the exercise wanted to be on hand to see the goings on, and to profit by the chance to observe the problem. As it turned out, the special duty for which Nate Hellman and I had been assigned was to take care of billeting, transporting, and making arrangements to ensure observers who were authorized would observe something. It was a two-week picnic.

While the rest of the regiment lay around the marshalling area, or aboard the landing craft, we were straightening out the few arrangements necessary to billet and feed several hundred visitors and get them to some vantage point on land or afloat from which they could observe the beach on which Tiger took place. I had about sixty officers, the lowest ranking of whom was a major, and he was by himself. The rest of them were colonels and generals. It looked like a headache, but I never heard a peep from them.

A captain from the Ninth Division whose name I forget had the job of taking care of Eisenhower. He went out on an LCI with Ike and generally had the time of his life. He said he never could figure out whether or not to stand at attention and parade rest all day—but that it must have been O.K. because he didn't receive any instructions on it. One pretty good incident occurred when he left the General. 7th Corps had issued a pamphlet on the Exercise, containing various orders, maps, reports, etc., which we had very strict orders to obtain from whomever they were issued to, and to return them after the exercise was finished. Remembering this order, the captain informed General Eisenhower that if the general was finished with the pamphlet, he would take care of it for him. Eisenhower crossed him up this time and answered that it was all right, that he would keep it.

The captain was definitely on the spot. He tried to tactfully explain that 7th Corps required all copies to be returned, hoping that the general would get the idea. Ike apparently got the idea, but he also had his own ideas. Very quietly he informed the captain that he would keep this copy. I can just about see him make the statement—turn his head up to the captain with the unspoken query, "Any questions?" The captain said he doesn't know exactly how he got off the special train but that he remembers saluting about four times and repeating, "Yes-sir" an unnecessary number of times.

Eisenhower spoke to the officers of our regiment during the early spring. He was a very impressive, dynamic looking person. He radiated his confidence—and we did have absolute confidence anyhow. I remember his remark—"We have a date in Paris, and I intend to keep it with you."

Incidentally, generals always caused a lot of headaches for the individuals who arranged their visit. To be presented to a general, all troops must, of course, be in ranks—and the idea usually was to keep a semblance of order throughout, even though the general wished the men to break ranks in order that he could gather them for a talk. We were most carefully arranged when Ike came through, with all dirty trench coats in back and rank strategically placed. The general immediately told us to break ranks, with the result that all dirty trench coats managed to acquire the front row and the rank was out in the cold behind where the younger officers climbed (and broke down) a wire fence to get a key spot.

Another time when Montgomery came through, the whole regiment turned out. The entire supply of engineer tape of the A&P platoon was used up to elaborately mark off the lines on which troops would stand formation on the large motor pool at Denbury Camp. In addition, a large semicircle was laid off, around which the troops were to move if the general wished to speak to them. Monty so wished—but the semicircle was too far away. So, he crowded them in around the jeep he used for his podium—and to break down the last sign of organization, requested that all helmets must come off so that he could get a look at the men. We almost landed on D-Day without engineer tape after that one. The A&P Platoon Leader blew his top when he saw his plans and tape shot.

Montgomery looked like a man buying cattle at an auction when he inspected troops. He has the coldest, bluest eyes I have ever seen, and they seem to look right into anything they rest upon.

He liked them brawny and big. If a commander was unlucky enough to have had small men assigned to his unit, he was apparently held directly responsible by Montgomery for not increasing their size by the necessary percentage. When he stopped to talk to a man, however, his attitude changed completely, and he was as personal as any general we saw. The men liked him spontaneously from this sole visit.

We had quite a dance for the men in F Company at Denbury about this time. Pulled in an orchestra and girls from all over the surrounding countryside. The Plymouth beer was weak but plentiful and the alcohol content added up. Schulzy, Lommler, Cook, and I got together and decided we would get the company tight, which was very easy. Just turned the taps and kept Capt. Fulton from getting worried. Sgt. Maines must have stolen the other mess halls blind for a week because there was more food around than I had seen in a long while.

It was a real rat race—Lommler eating fire and Schulzy figuring out how to meet a little blonde on the way home. The dance floor is off limits to officers at an enlisted men's dance, although they always come to their own unit's dances. In England, the officers had to ride the trucks which took the girls home, however. That made fast work, if any, a necessity. Cookie was always the pappy to the company and had his usual following. I remember that I must have put up a good front prior to this time, because Ivy in my platoon, came up with a new look of respect in his eye, and said, "Gee, Lieutenant, I didn't know you liked to drink beer so well and chase women, too. I never saw you let your hair down before."

Honor, Duty, Country, and the Officer's Guide. That's me.

Exercise Tiger at Slapton Sands claimed the lives of almost one thousand American soldiers and sailors. On April 27, 1944, during the Exercise Tiger, German S-boats were able to sink two landing ships and seriously damage a third loaded with soldiers, tanks, and equipment. Failures in communication, the freezing water, and lack of training in using life vests contributed to the disaster. Additional deaths were caused by friendly fire during naval bombardment simulations. The tragedy was carefully analyzed, and lessons learned, but would not be formally acknowledged by the U.S. and British army until 40 years after the war.

There was no way to know if my father's silence on the subject was intentional—he was certainly there and involved with the observation teams.

16. D-Day

JUNE 6, 1944

On June 6, 1944, the New York *Times* announced, "ALLIED ARMIES LAND IN FRANCE IN THE HAVRE-CHERBOURG AREA; GREAT INVASION IS UNDER WAY."

Kathryn Sisson Phillips sent a telegram to her brother Warren and Margaret Sisson, Lt. John Sisson's parents, that morning:

With you every moment of this day in prayer for John's safety.[46]

I couldn't begin to imagine how my Aunt Kathryn felt reading those headlines in her apartment on Fifth Avenue in New York. Kathryn and her brother Warren had lost their beloved younger brother Horace to World War I. They knew all too well that war can be cruel and fickle. As for my Grandmother Sisson, I believe that only mothers of combat soldiers could know her agony.

```
About May 15th we packed up one day and said
farewell to Denbury.
   Our new home turned out to be pyramidal tents
near Plymouth. This was known as a marshalling
area and the next stop was France. Everyone knew
this was it, although we had not as yet received
our orders. Hikes, games, and physical exer-
```

cise were the order of the day. Equipment was stripped down to necessities—practically what one could carry on his back. Then they began to add new equipment here and there.

Hal Simon started his famous S-4 career in earnest about here. There is probably a warehouse officer wandering about somewhere even yet, wondering just how half the equipment in his charge disappeared—with his permission—shortly before D-Day. The 2nd Battalion was extremely well equipped. With both S-4's and equipment.

Our move to the marshalling area was of the highest secret type. We were to disappear in the night like a whiff of smoke. When five girls from Newton Abbot, which we had left some 50 miles behind, were found hiding in the woods, waiting for their dates to climb the fence and see them, various and sundry parties began to flip their lids. CID,

MIS, Corps, Army, and Division liaisons were running all over the place. A real field day. A major was accused of letting out the secret, but the name could not be obtained. Majors Edwards and Dowdy spent the next three days turning the knife in each other's back.

Finally, we pulled out—after an amazingly detailed briefing as to our exact landing position and the layout of the whole Cherbourg peninsula. We boarded an LCI, as expected, and settled down to wait. The sailors were so used to dry runs—as the problems were called—that they sat around unimpressed as we boarded. They had been told nothing. However, a crap game started immediately—naturally—and when one of the navy boys idly inquired what the stuff was they were using for money, it was started.

We had been issued invasion currency, plainly marked, "France." When the sailor spotted that he gulped twice, looked up at the owner of the bill, looked at it again, dropped it like a red-hot coal, gave a loud "Yipe" and dove down the nearest hatch to spread the happy news. There were no more bored looks, and the navy came boiling up out of the hold like a swarm of smoked out bees. The big show was on. For three days we sat around, waiting for the word, "Go." Each day the take-off was postponed.

Finally, on the night of June 5th, we were out to sea with England out of sight. We glimpsed the other convoys from time to time, and the channel was literally afloat with ships. As we approached the ship rendezvous area on a highway of water marked out by buoys for miles on end—we saw more. It was about this time that the USS Battleship Texas' batteries were not getting a chance to cool. We could see Omaha beach off to our left—our own Utah beach dead ahead. Battleships were sending up great flashes of smoke and flame as they threw their shells shoreward. Rocket boats disappeared in a cloud of smoke as they discharged their deck full of rocket launchers.

The sky was full of buzzing planes, and we

could see them dive down to strafe the beach, still several miles away. But there was an air of unreality about it. We had prepared for this for so long and had run through the same thing so many times before, that we could see nothing different from Slapton Sands. Even as our LCI edged in toward the beach and we could see sudden geysers spout up around other craft not far from us, it seemed unreal. You couldn't see the enemy.

Everything was going just as planned. Twenty minutes ago, I was shaving in the latrine in my underwear. You could reach out and get hot coffee. The boat was moving the same as it had always moved. Sgt. Conly, technical sergeant for my platoon, grinned back as he always did before we got our drawers wet on a problem.

The only change was that we didn't have to neatly pile our life preservers when we hit the beach. This time the order was "just drop 'em and move like hell." Someone thought combat was OK because of this point. As many men were talking casually with no particular attention for the shore as were watching the proceedings landward.

An LCM—open scow-type with a front ramp which dropped—pulled up alongside. Our LCI was first in line for our battalion: F Company spearheaded the 2nd Battalion, and my third platoon spearheaded the company. We clambered over the side and into the LCM. The LCI couldn't get in, because the tide was dropping too fast, and the water was too full of mines for a boat its size. Looking through the slot in the ramp as it was raised into its upright position was like watching a Hollywood camera close-up on a subject from far off. The shore filled more and more of the slot—and features began to stand out.

The coxswain yelled once, and we looked over the side and saw a mine we had narrowly missed. Things began to look a little different. A boat off to our right was bracketed by sudden geysers. Some G.I. looked uncomprehendingly for a minute—then as if getting the picture, said, "Why, that stuff's Jerry's. Those b------ are trying to kill us sure as hell." Up until then, some of the men had wondered how they would bring themselves to use the weapons they had as coldly as they were calculated to be used. What would it be like to shoot at someone? Would I be too scared to do anything? Would I do something heroic and brave?

Suddenly there were no more questions. Only a G.I. saying, "those b------- are trying to kill us sure as hell." That is all that is needed to make a soldier fight. That is what makes him do brave things or cowardly things. Just that one phrase. There is no motive or thought behind a trigger squeeze. Sometimes it gets a little more positive, and the G.I. says, "I'll kill those b-------" but that's usually after a long time at it with too many things to remember and too few which can be forgotten. Besides "those b------ are trying to kill us" is always true. I have yet to meet a soldier, I believe, who would go on killing Germans if they stopped trying to kill him.

Our boat hit the beach at about 0930 or 1000 hours on D-Day, some two or three hours after H-Hour. By this time, the beach was beginning to crowd up—but we luckily had none of the direct fire, at that time, which threatened Omaha Beach where the First and Twenty-Ninth Infantry Divisions landed. A few 88's were landing not far away—but we were able to proceed according to plan and get off the beach and inland as rapidly as possible. The plan called for us to land directly into the teeth of three strong points and knock them out if they had not already been taken care of.

Good luck, I think, caused the beaches to actually be set up about 1,500 yards south of the designated spot. Accordingly, we were out of range of the strong points and were able to proceed inland with little or no opposition. The road across an inundated area, just over the beach, was jammed with every sort of soldier and vehicle imaginable, and why Jerry didn't zero in on it is more than I will ever know.

After we oriented ourselves, we struck inland, and hit a road which led to our designated rendezvous spot. Captain Fulton was with me, and he wanted to push ahead. So did I. The farther we got without a fight, the cheaper in the long run. We posted at least four guides behind us to direct the remainder of the battalion—and of F Company and struck off down the road with about fifteen or twenty men, including George Caldwell who was our artillery forward observer. We marched steadily for about two hours and reached our rendezvous but had completely lost contact with any other troops.

We contacted the paratroopers who did a magnificent job of clearing the way for us. Each time we would see them—we were the first infantry troops up to this area—they would spot our ivy leaf patches, and grin, "Man, I was never gladder to see anyone in my life." It sounded as though they were done fighting and would let us relieve them when they said it. But each time, they would turn north again, and go ahead to knock off another bunch of Jerries or take a little town from him. The fifteen men I had would have been in pretty sad shape if it hadn't been for them. It amounted to only about a third of my platoon—the remainder of which it turned out were following us at about a mile.

Capt. Fulton went back to contact the rest of the company and about this time Caldwell decided to shoot the first round for his battalion. So, he started looking around for an observation post. A housetop looked good, and my protests that the town just beyond was still in German hands didn't affect him in the slightest. So, if the guy had to be stubborn, I would help him. We tugged and pushed and finally got him atop the house. Then he decided he couldn't see enough, but a tree nearby was just the thing. And it would have been if German snipers hadn't been firing from any and all trees all morning. No American with good sense would get that far off the ground. At least that is what the Jerry in the field a few hundred yards away must have thought. He pulled up and began to fire at George in his tree.

I now believe the Darwinian theory of monkey to man, completely. Caldwell did everything but use his tail in coming down. Weissmuller would have cried out in envy at his interpretation of Tarzan. And each time he dropped lower, a bullet cracked over his head, chasing him down. Finally, the Joe firing at him decided to lower the boom and drop one in low. It caught George in the side of the foot. No great damage, but he was definitely out of action. He was really sore. Said he could see the whole peninsula from where he was and could have wiped out Germans if they stuck their heads out of a window in Cherbourg. We wrapped him up and left him with a couple of German wounded and a paratrooper and pushed on. The medics would find him. We hoped. They did.

We pushed on by the map across country. Fortunately, the fiasco at Slapton Sands led to some intensive training on hedgerow tactics so we were prepared for what we found. It was all according to plan and was going too good to be true. We saw a few people—and they were apparently typical Normandy characters. They didn't seem very liberated—in fact we had only one smile from the dozen or so we saw. We had waited where George was hit, until the remainder of the company came up—and from there we pushed out together. The rest of the battalion was out for lunch as far as we were concerned, though. We didn't see them for the next twenty-four hours.

Darkness came very late—so we had pushed up the peninsula toward a town called Azeville by dark. Weingart had the first squad and, as usual was leading them when I caught up to him just as he came to a hedge row running diagonally across our front. A road was on the other side and just as we were looking over it we heard the popping of some sort of vehicle coming down the road. We hardly got a look at it before Servat let go with his BAR and things began to happen in a hurry. Something exploded, and I figured Jerry had some artillery on us.

We crouched down behind the hedge row—and as we did so, heard a car slide to a halt directly across the hedge row. Peering over, I just made out a grey-green uniform diving into the ditch along the road. Weingart and I looked at each other and both reached for a grenade at the same time. At that precise second, all hell broke loose around there, and shells, shrapnel, and fire started shooting in all directions. "Weingart" I said, "let's get the hell out of here."

We hit a shallow ditch about fifteen yards away as one man. As it turned out, we had hit a German ammunition truck, and it went along merrily for about fifteen minutes until another vehicle came along. Servat got this one too. It so happened that the second vehicle was towing an "88" and that it stopped exactly in the center of a crossroad, forming a perfect roadblock. We let the fire burn out and rounded up the remaining Jerries. No one was so much as scratched.

Buck Camper was nervous as a cat after he discovered that he had been standing in the road for a full minute within ten feet of a hiding Kraut whose rifle was loaded and ready. Les Kumouwski had a bullet so close to the seat of his pants he thought he was hit—but it didn't touch him. My carbine jammed just as a Kraut jumped up from the ditch about four feet from me and swung his rifle around to fire. He fired once, but not again. Mead was around as usual and took good care of him. We organized the company around the crossroad and dug in for the night.

17. A Battery and a Chateau

JUNE 7 – 10, 1944

The initial objective of the 4th Infantry Division after landing was to secure two of the most powerful German coastal forts in Normandy, the artillery batteries at Azeville and Crisbecq. They threatened the beaches and shipping lanes and lay between Utah Beach and the port of Cherbourg, which was the 4th's primary objective.

The Azeville Battery would be the Second Battalion and F Company's next challenge.

Meanwhile, I tried to imagine how each soldier in my father's platoon felt as they awoke on June 7 in a maze of sturdy hedgerows with no real idea of where they were. The hedgerow was a fence that was half earth and half hedge. Brambles, vines, and trees interwoven to reach heights from three to over ten feet. Farmers utilized them as fencing, so they literally broke the Normandy fields into walled enclosures, well defended by the outnumbered German army.

```
In the morning, we had about five vehicles and
the "88" stretched out along the road.
    We were still just below Azeville. Jim Beam
took out a patrol and came back with a Jerry burp
```

gun and a story of a couple of close calls. In one, he lay in the hedges of a row along a road while a five-man German patrol marched by him. This was where we picked up Smitty, the German first sergeant who carried our [SCR] 300 radio for a Pfc. for the rest of the day. Fred Mason came back with a story of having been chased for about a mile through the fields by a couple of Jerries. That was the sort of fighting it was. No one knew where anyone else was—friendly or enemy.

We moved around to the left of Azeville and went into position. Couldn't see anything so Capt. Fulton told me to take a patrol and find the battalion, which by now should have been up on our right somewhere. We went ready for anything, so the patrol was of good size. We took off and within an hour had located the battalion CP. They didn't know where we were, and Maj. Edwards was rapidly getting jerked off. We pulled into the battalion left flank, and my platoon set up flank security around a farmhouse. We got our first crack at fresh milk since we had left the States and some of the first eggs in a long time. The farmer tried to marry his daughter off to Mead, I think. He figured he was quite a lover.

One of the strangest things was the first time the supply vehicles came up. It seemed almost impossible that they could actually get there. But there they were. A beautiful sight. The Tank

Destroyers—artillery on wheels—rolled up that night, too. A lovely sight.

Later that night, Carter and Erikson came in with five Jerries who had come parading down the road. They had a small arsenal on them but had dropped it all when Carter snapped the safety on his rifle. They must have been part of the counter-attack that we received later that night. There were German automatic weapons firing everywhere—and things were looking bad when our mortars suddenly opened up. They walked back and forth across our front and then began to chase the Jerries back to their holes.

The next day, we pushed up directly under Azeville, where there were four casemated 155 rifles in battery. I got orders to go up a wide-open hill with my platoon and take the battery. An order to pull back and prepare for a counter-attack postponed that one. It took a battalion to take it the next day—and they approached by an entirely different route. Mead was pulling his usual stuff. He took the sniper's rifle away from our sniper and started a personal battle, at 200 yards, with a 20 mm or 40 mm anti-aircraft gun that was in position on top of one of the pillboxes. He claimed he got one, too—and I wouldn't doubt it.

Artillery observers came up and we fired 105's on the position—even had the cruiser USS Lexington firing for a while. One of the Lexington's shells made a direct hit on the pillbox, and before the smoke cleared we figured there would be a beautiful pile of rubble left. When the smoke cleared, there was not only no rubble, but the paint wasn't even scratched as far as we could make out.

We pulled back to the farmhouse that night and got a half-hearted counter-attack. There were no lines on the peninsula then. A perimeter defense was thrown up—and this was usually not too tight. Schulzy awoke at daybreak one morning to find a Jerry standing above his hole with rifle in hand. Schulz coughed and shifted enough to allow him to grab his carbine and came up spraying lead. The Jerry took off, and never was hit. Schulzy was about as close to the center of the battalion area as you could get!

On the fifth or sixth day we attacked the Chateau Fontenay, or some name similar to that. Jerry got on our flank and a sniper crept in to within thirty yards of me and got me too easily. This started a long trip back through channels. I think I used every means of transportation around. I walked the first mile to the aid station, rode the next trip in a jeep ambulance, rode back to the field hospital in an ambulance. A "Duck" rode me out to an LST that was our hospital ship, then an ambulance took us to a hospital train. I was cheated. No airplane ride.

Nothing in the army has impressed me more than the medical set-up. Those guys give you all the confidence in the world in them—and do such a good job that you don't worry much, anyhow. All this was only five or six days after D-Day, and they hadn't had a picnic getting set up. It all went off like clockwork though. Some joker by the name of Snead kept telling me I'd get sick when he put me to sleep, but he had me laughing so hard that I didn't know what hit me. It took several days to make it back to England—traveling the whole time on a stretcher was a big pain in the neck. Finally ended up at the 185th General Hospital near Taunton.

18. The Men of F Company

JANUARY 1945

I had begun to realize that the description of strife and combat and heroics that was so evident in the histories of war that I had read wasn't going to be found—in the traditional sense—in my father's story. This surprised and perplexed me. No mention of the stormy seas the men braved to land the morning of June 6, no mention of the emotions surrounding surviving the landing and being among the first Allies to force their way into an occupied continent, no mention of the swampy defensive inundated areas between Utah Beach and Saint-Germain-de-Varreville.

My father's story seemed more about soldiers doing their job: the paratroopers that landed the night before; Carl Servat, their BAR man, skillfully using the gun he was trained on; Sgt. Al Mead's quick thinking; squad leader Sgt. Bill Weingart's leadership; Captain Fulton's ability to inspire respect; and many more. Dad almost seemed to be a casual observer. There were two sentences about the battle at the Chateau. Not much more than he had already told his family about the bullet he carried throughout his life.

Although the impact of the war was not evident in my father's narrative, I could sense it in the following tribute to his men, whose names I have highlighted. This section is the only place where he stepped outside the chronological narrative to write as a survivor and one with knowledge of the fate of many of the men as of January 1945. I feel a deep sadness every time I read this section. There was one death he didn't know about until many years later. There were probably more, but I only know of this one.

A great many things ended when I started that trip back from the chateau.

Most importantly, ten months of training and learning to know the men and officers of F Company. Some were still there, others weren't. But we had begun to break up now. And with the beginning, was the realization that it would continue. Old names and faces would gradually disappear for one reason or another. New ones would come and go. Once you could put your finger on F Company. Now many hands would be needed to cover what was meant by that same term. I would always be able to put my finger on my platoon, though. The new faces would only represent the old ones they replaced.

Jim Conly: Tech Sergeant and platoon sergeant. Tall, fair, a Carolinian with an engaging smile and a rugged set of fists. He was wounded around Cherbourg—but I'll always remember him saying, "Ah laowks that." He always kidded me by telling me he was going to take over my sleeping bag when I was hit. But it came through to me in England—so he must have slipped up. Or maybe the screaming meemies kept him too busy.

Al Mead: was my favorite character in the Army. A platoon guide has a more or less administrative position in a platoon; keeping records, reporting to the company CP during combat, etc. but Mead was undoubtedly the least administrative individual in the outfit. His position, so says the manual, is in the rear of the platoon. I can never remember him more than fifty feet behind me—and usually ahead of me. And I was finding out that they used the word precisely when they said a lieutenant was a platoon "leader." Mead, as I said before, probably saved my life the night of D-Day when the Jerry in the ditch decided to show some fight. The next day when the company had pulled out of a position after being shelled at point blank range by 88^{1}s, he decided he would go back to get his pack where he had left it—so we organized a patrol and waltzed out a mile or so to get it and some others. The next day he picked out the 20 mm anti-aircraft gun for a personal duel with a rifle at 600 yards.

He was hit at the chateau, too. A machine gun was spraying the side of a building near a gate—but he thought he might be able to get a shot at it—so he walked right into the burst and caught one in the hip. Conly and Mead had it in for each other when I joined the outfit.

They were both definite individuals, so there was nothing surprising about that. Mead, for my money, was the best non-com I've ever had anything to do with in the army. His mother lived in Cincinnati, and he was devoted to her. I used to get a kick from censoring his letters. He had a half dozen girls, all of whom he kept on the hook with a beautiful line. Identical for each one. Tall, lean, dark hair, and Latin appearing, he was strictly soldier, with a crack always ready on his lips.

Bill Weingart: was a kid of twenty or so from New Brunswick, N.J. Mead took over the first squad while we were in Fort Jackson and kept it until shortly before the invasion. In England, Weingart was the natural choice, but due to some misunderstanding with Capt. Fulton, I hadn't seen to it that he had it officially any earlier. Weingart was strictly on the ball and as sharp as a tack. In combat he turned out to be far and away my best squad leader. The day the platoon was supposed to take the Azeville guns, his own rifle jammed while he was crawling up the hill directly into the face of some 20 mm fire. He just curled up, disassembled his rifle, extracted the ruptured cartridge, put it back together again and kept going until the attack was called back, shortly afterward. If I ever gave him an order, he was either already doing that very thing, or getting ready to do so. He was up for a commission when he was wounded later in the Cherbourg campaign.

Paul Anish: was a T/5—a rating left over from the 4th's motorized days—he took a crack at being first scout and did well enough. When I saw him last, in Germany, he had had a squad and was busted for some reason.

Rastus Durham: better known as Bull, was a self-made Georgia boy. With an accent a yard thick, he was a little hard to understand at times. But he had taught himself to read and write while in the army and was a staff sergeant in charge of the second squad. I remember the day we were training with toggle ropes and working only from memory and logic. He lined up the correct number of men and, by the numbers, put them through the construction of a twenty-five-foot rope bridge. Somewhere along the line a shell scared the hell out of him, and he made no bones about it. He's back at regimental service company now and says it's much better there.

Sgt. Fitzgerald: was buck sergeant in the second squad. He was shrewd, red-headed, and talkative. He was around for a lot of it after I was wounded, but is back at regiment now, too.

Bob Pollock: was too good to be a buck sergeant, but never got a break until combat. It's no break to last a long time there—but he ended up as first sergeant, and he was easily capable of it. He was very devoted to his wife, Bee—probably the steadiest fellow in the platoon, even if he was the hottest crap shooter.

Ozzie Wirtzberger: was an easy-going Dutchman with a wonderful sense of humor. He was from Frankfurt, Kentucky and loved to recall the better and bigger beers of the Ohio River valley. Even the invasion had a hard time breaking up the pinochle game he presided over. At the Chateau, he led the attack. A sniper caught him across the top of the head. He's buried near Sainte-Mere-Eglise, with Carter, who was hit twenty feet from him.

Ralph Carter: was a private in the rear rank, but he was morale. If any statistics on the Detroit Tigers or University of Michigan's football team escaped him, they failed to exist. To him anything was worth arguing about, and his lack of information or logic on the subject never bothered his volume. He just mowed 'em down. About five foot four, he was a perfect likeness of the Stars & Stripes Hubert. Jenkins and I dressed his wound at the Chateau and Jenkins stayed by him for over an hour. We couldn't get litter bearers up for either him or Wirtzberger.

Laurel Pierce: was one of the best men in the

platoon, even though he was only a private. The first contact I had with him was when he signed his mail, Corporal. When I chewed him out and restricted him for a weekend, I liked him so well, that I lifted the restriction before it began and made him one of my runners. Capt. Fulton made me mad as hell when he refused to give him a Pfc. (Private First Class) rating, and then snapped him up as his own runner. He and Jim Beam got together—Jim liked him as well as I did—I have heard they were inseparable until Jim was hit. Pierce was hit at about the same time as Beam, if not by the same shell. They didn't tell Beam that Pierce was killed. That was down around St. Lo, after Cherbourg had been taken. Pierce could jitterbug the skirt off any gal in England and would have gone far if he had given a hang. He liked it better his way. He and Chuck Phillips were inseparable buddies. Phillips took it pretty hard when he found out about Pierce while we were at the rehabilitation hospital in England.

Les Kamouwski: was a funny little guy with a Cheshire grin that belied his cynicism. He hated the army—but bless him, he never showed it. He must have been pretty well educated and, at any rate, had plenty of intelligence. I thought he was our first Purple Heart member when his pants were singed on D-Day, but nothing came of it. That crooked little grin was there when he said,

"Lieutenant, I think I've been hit." I believe he is OK yet. Didn't see him around in Germany.

Bob Ivy: made up in pure animal spirit what he lacked in brains. He could bounce around like a three-year-old and never get tired. He had seen a lot of action and was still going strong when I left the battalion and started home on this leave in December.

Red Williams: was Ivy's buddy. He used to buck for guard regularly—with everyone from Conly on down shining his brass and shoes. He was as good a rifleman as you could ask for. Ivy tells me he was killed in France.

Erickson: was a quiet guy, and I didn't know much about him. He and Carter brought in five Jerry noncoms from the outpost one night, and they were more surprised than anyone else. The next night I was checking their outpost just as Beam was coming back down the road with a patrol. It was at the same place that the Krauts had been picked up the night before. We halted them, asked for the password, and got no response. All of us had pulled the triggers of our guns to just about the release point when Jim said, "It's Lieutenant Beam." That was too close for me.

Briggs: came from Michigan. He must have been

over age when the army got him—but never was there a more loyal and obedient soldier than he. Although he was old—I don't know how old, but well over 30—he was as strong as an ox and would rather fall in his tracks than complain. I hated to have to do it, but he was transferred to some non-combat part of the regiment just before the invasion. It broke his heart to leave the platoon, and it took everything I had to tell him he had to go.

Charlie Wilson: was a quiet little guy who was always ready. He wasn't with the platoon long enough for me to get to know him, but he was still with F Company when I left them in Germany in December.

Luke Montague: is a big moose of a guy from Philadelphia. He came from West Philly, and **Carmen Colorrusso**: came from East Philly. They were the two roughest guys around, and their favorite amusement was throwing each other on the ground and jumping on the other. Montague wrote me while I was in the hospital—"Well, sir, how are you. As for me, I am fine." He's on this leave too, having been wounded at the Chateau, again during the St. Lo breakthrough around the end of July, and once more lightly in Germany. He was sweating this trip out, thinking he wouldn't make it for a while, but it worked out. Colorrusso wasn't around while I was back with the battalion in October and November—but he is well I hear.

John Schaming: was a little guy, but with a BAR he became bigger. He could fire one of those things as though it were part of him. He's along with us, too, on our leave home. I don't know where he was injured.

Lou Petrocelli: was an Italian kid who wanted only to get back to his wife, whom he had married shortly before we came overseas. A Greenwich Village kid, he greeted me as a long-lost brother when he found I knew of Bill Bertolloti's in the Village. A glass of wine with Louie at Bertolloti's when this is all over…

Louis Nardiello: was company barber, and third squad operator. He caught some shrapnel in his leg and has been put on limited service. For a skinny guy he could carry more, farther than I would ever have believed. He didn't think much of war, either. Used to give me a fine haircut, in fact I owe him for one in the marshalling area yet. He was from around Bear Mountain, N.Y., and I hope he gets his troubles back home straightened out. He really had them according to Louis.

Fred Mason: was strictly character material. Before we came overseas, he was in more scrapes, was AWOL more times, had more court martials and served more time in the brig than anyone in the company. One day on a hike around Denbury he collared me and informed me that he hadn't had the right breaks and proceeded to do a lot of moaning. For some reason, I took one look at the guy, and decided he was right. To the third platoon he came—with the understanding that it was soldier-or-else. He did it, too. He made orderly on guard at least four times running and tried hard as hell to stay out of trouble. Some amateur boxing in and out of the army made this difficult on occasion. I figured he would make the best first scout I had. He did. He never said uncle, and never stopped going. I couldn't find out where he is now, but I hope he hasn't forgotten that he'll get his breaks if he makes them.

Jim Barrett: was assistant company clerk while we were in garrison, and strictly a New York character. As Irish as they come, with a large supply of the old blarney.

Carl Servat: was 3rd Squad BAR man and would have made a good noncom. He and Pollock had many a happy night on the blanket with a set of dice. His BAR got our first crack at the Jerry, and never missed after that.

There are lots more—**Joe Smith, Suozzo, Raab, Dick K.,** etc. Of the bunch there wasn't a lemon.

To me, that will always be the third platoon. They were a wonderful bunch.

The officers, **Capt. Fulton, Jim Beam, Jerry Claing, Bill Schulz, Cookie (Lt. Cook)**—were all tops. Beam lives in Lincoln, Illinois and Schulzy, in Syracuse, N.Y. They're both back home to stay now, having been seriously wounded in action. Jerry got his captaincy and was wounded when he fired a rifle grenade with a live round instead of a blank up in Germany. Whatta a guy. He's from Hartford. Capt. Fulton and Cookie live in France now.

This last statement confused me until I realized my father was saying they were buried there.

19. Too Good to Be True

SUMMER 1944

Regret to inform you your son Second Lieutenant John F. Sisson was seriously wounded in action ten June in France. Letter containing present address follows.

Western Union telegram from the Adjutant General to Warren and Margaret Sisson[47]

Among my father's papers were a series of letters home in June and July where he downplayed his wound and the violence of the ongoing battle in Normandy, to the point that my grandfather, no fool, eventually contacted Washington to try to understand the truth of what was going on. The answer was a standard Army issue letter that said nothing. I imagine there were many similar exchanges between sons and worried parents. We were told growing up that Dad still carried the bullet because it lodged in his abdomen in one of the few cavities where a bullet can penetrate and not kill you. He said Snead remarked in a jovial way when they first looked at him, "Man, what are you doing alive?"

Back in England. It was too good to be true.
Sheets on the bed. A little brunette nurse from Sandusky, Ohio had the big pleasure of

shaving and bathing me for the first time. Breakfast in bed, first square meal in twelve days after the hospital train got me to Taunton, Ward O-1 at the 185th. Bedpans finally got me on my feet. Nature calls ought to be heeded more than once in sixteen days. And I hate enemas.

Our non-stop hearts game started shortly after Verne Summer—Capt., FA, 90th Division, Rush Springs, Oklahoma, and the man that damn near killed me with his antics—and I got out of bed. The pure relaxation of being out of the lines made us so slap happy that we would laugh at anything. We did.

My family blew their collective tops when they thought I was too casual about getting hit. I blew mine, too. It hasn't settled yet. Dad wired quote: "Insist no more combat." I filed this under S and referred it to Eisenhower.

More mail than I had ever seen in my life. About 52 letters caught up with me that first day.

About the time I was getting around some, they threw some more stitches in me and kept me in bed for a couple of days. After about a month, I got a pass into town. Not feeling very active, though. Rapidly began to feel better. Verne and I were discharged together and assigned to the 77th Rehabilitation Hospital, Bromsgrove. It is near Birmingham and Moseley Village.

We started out one fine morning—our first time out of captivity since we went into the marshalling area about May 15th. It was August the 1st. England was the immovable object, and we were the irresistible force. What a collision! The train was late in Taunton, so we walked down the street toward town. A jeweler was outside his shop on a ladder washing windows, when Verne pulled up under him, and said, in that Oklahoma drawl, "John, let's go in here and spend a couple hundred pounds." I pushed Verne out of the way in the nick of time as the jeweler came down off the ladder and into the shop like well-greased lightning—rocket propelled. Verne would have been trampled in the rush.

We stuck our necks in and found the old geezer grinning from ear to ear and wiping off his counter with a chamois. After letting him sell us the shop—mentally—we bought a thirty-five-cent watch strap, and left him with the definite idea that his shop carried only second class, bourgeoise goods.

The train got us to Bristol in mid-afternoon. We were out of cigarettes, so I suggested we go to the Red Cross and see what luck we would have. When Hazel Tarbutton, ex-Ohio State buddy, greeted us at the door, I knew we could do no wrong this trip. We got a carton plus a pocketful of chocolate.

The restaurant we wanted was closed, but another was recommended highly. We didn't have the first idea of how to get there but had an address. We spotted a Ford staff car with an

English WAC driving it parked in front of a hotel nearby. That looked like a fine solution, so we piled in and informed the driver that, "The colonel told us to have you drive us out to the Black Swan." "Right-ho, sir," and off we went. I sweated out every colonel I saw for a week afterward, but our luck held.

We taught four or five kids to quit asking for pennies and get into the big time. After some coaching, they got the idea and were working on the line, "Got any dirty old pound notes, chum?" A couple of scotches got us well lit, after our long and just finished exemplary life at the hospital. We wound up at the train station in the middle of the night and went to sleep in a coach we found handy. We woke up when we were five miles out of town. Everything looked fine, but we didn't know in what direction we were going. We got to Gloucester—I still don't know where it is—and changed for Birmingham.

Verne had to explain our Purple Hearts to everyone. The invasion being new, we were the toast of our six passenger third-class compartment, which had ten people and two children at the time. Verne and I weren't the latter, although you could argue the point.

We hit Birmingham about teatime and Verne found the Grand Hotel somehow. He was out waiting for Peggy, a girl he knew and the main reason for our trip, when I ran into Joe Sampson, ex-8th Infantry Lieutenant. There were only about a hundred of Birmingham's typical conservatives around sipping their tea when he roared in and greeted me with up-raised cane and a foghorn voice you could hear halfway to London. I had never seen the guy before, but he soon convinced me we were old buddies from LSU.

Things seemed fine enough until he started screaming a bunch of derogatory remarks about Birmingham, English beer, the whiskey shortage, and generally informing me of the promiscuity of all women thereabouts and the fact that the hospital we were going to was a hotel and playground. Specifically, he displayed a colorful and varied vocabulary to an audience of Englishmen, most of whom were close to drowning amid the spluttering in their tea. I was so tired I couldn't do anything but sit back and laugh. So I did. He excused himself to everyone in earshot (three blocks) when he spotted a likely looking blonde, explaining that he had to go "make a girl." His introduction seemed to work, too. Anyhow, she knew exactly where she stood.

After Verne had seen Peggy, we took off for the hospital—which was indeed a playground—and slept for twenty-four hours.

It was supposed to be a rehabilitation center, but with nightly passes most of the patients probably left in worse shape than when they arrived. It was a great idea, though. As a matter of fact, they had a very complete and effective program to gradually work a man back into shape.

Verne got on a bus, going into Birmingham with me one day, but when I turned around, he wasn't there anymore. A short look around found him in a corner, with an English girl backed farther into same, obviously trying to figure out just what the devil was going on. And so, we met Margaret. Verne came home with reports of a friend. And so, we met Val.

Before I was assigned to the 307th Rehabilitation Hospital, we all had time for a picnic—rare stuff in England those days—a swim at Droitwich, and a trip to the Green Dragon in Sambourne. Verne and I spent approximately every night commuting to and from Moseley.

They wanted some lieutenants at the 307th—an enlisted man's rehabilitation hospital—so I signed up and was picked. Had "U" Company for the weeks I was there. Outside of the fact that every G.I. in the place knew he was shortly on his way back to the front, and generally didn't give a damn, it was a rosy spot.

The rosiest part was my first day, while I was inspecting the barracks of the company to see just what I had on my hands. The voice in the rear of the second barracks I stepped into was unmistakable, but I couldn't believe my own ears. Mead, no less. He was thoroughly anglicized, now. His girl in Leamington had shown him the light.

While I was there, I saw Sgt. Phillips, Ratterman, Bailey, Gillenwater, Sgt. Dowdy, Joe Smith, and a few others—all of F Company. Between weekend trips to Birmingham, and weeks of sweating out the lieutenant who had our rehab battalion, things really ran unevenly. It was a screwball institution but fun. Paid 10 pounds for a bicycle I couldn't get rid of when I left. I'll bet Hank Allison still has it.

The big call for infantrymen came out in September. As I was all well now, I was put on orders. Col. Stinchfield was our C.O., a very remarkable and capable guy. On the morning of September 25th, Lamberson, Kelly, Sutton, and a few other members of the Leamington slap happies congregated at the officer's mess of the 307th.

The party was over. Nervous in the service began again. Big laughs, painful silences, irresponsible combat psychology—it all came flooding back in a hurry. Three days ago, any one of these officers, all lieutenants, would have chewed, restricted, and fumed if an enlisted man failed to salute properly or wear his uniform correctly. Now? Everyone else was chairborne—rear echelon—SOS. Salute? I heard of it, once, but to mention it in public was almost a sin. I'm a combat Joe, now. I'm rugged. Let's have a drink and have some fun. Whistle at every gal. Cock your overseas cap at the Rest position. Never pass a pub. Hell, man, the 4th Division is the best damn outfit in the U.S. Army.

These guys back here in England are a bunch of sob sisters and politicians. I want to get back to my outfit. Back home.

Our bus from Leamington to Lichfield, home of the 10th Replacement Depot, was one of these English bus affairs. Fifteen feet high, square sides, driver in his little cab at the right front. Piled high with

baggage, bed rolls, val-pacs, musette bags, boxes, duffle bags, etc., it looked more like a junk wagon with windows.

This was about September 20, 1944. Lamberson and I, who had fought it out as rival and adjoining company commanders at the 307th now became big buddies. He went to Pheasey Farms and had it good. Apartment houses, private bathroom, suite of rooms, buses to Birmingham outside his door. I was at Lichfield—ten miles out and a member of the Killian health resort. The word enroute was: "You're in like Flynn." All day long, "You're in like Flynn." When I went down to Pheasey to see Lamberson, there was the guy, Flynn, himself. A big Irish CIC lieutenant. The Flynn business got overworked from there on out.

It took about a week to get on the way from Lichfield. In the meantime, progressed one of the greatest rat races known to man. Birmingham was assaulted daily. The Hungry Man pub was stormed, stamped, and approved. Lamberson and Sisson spent four days chasing shadows and breaking the quiet at 117 Anderton where my friend Val and her roommate Jackie, for two, were behind us to the end. Getting taxis or buses from Moseley to the city in the middle of the night was not too much of a job. Many ten-shilling notes dangled as bait for nocturnal hack drivers to return for us.

    At Lichfield, I received my assignment for the return to France. Finally, we were ready. Footlockers packed and sent to Liverpool for storage. Extra clothes thrown away. Musette bags crammed. We got a train in the middle of the night from Lichfield and were off to Southampton.

20. Return to the Front

FALL – WINTER 1944

As I edited this chapter, I found myself wishing that I knew what had happened to Dad's buddies Kelly, Flynn and Lamberson as well as Verne Summer. In 1991, George Bridgeman, Dad's neighbor in Galion, Ohio, sent word to the 22nd Infantry Association and a very nice note to my family when he received news Dad had died. My mother said they had stayed in touch. I never did find out if Kelly, Flynn, and Lamberson had first names, and Val and Jackie, well, they never had last names. I hope they all prospered after the war.

In this chapter, my father returns to a different European Theater. Normandy was secured, Paris liberated, and his 4th Infantry Division was knocking on Germany's doorstep. None of them knew the hell that was waiting for them that winter. I had yet to find any clues to help me with the mystery of the *Mein Kampf* story. But I had gotten to know and care for F Company and the brave 22nd Infantry Regiment. That had been well worth the journey.

```
Connections were good and we boarded the channel
boat per schedule around September 30th.
    Lots of Frenchmen aboard were going home for
```

the first time in four or five years. One girl I talked to had been in the underground movement. Told of how she escaped France in a hearse as the body. All the adventures aren't in Colliers and Hemingway. The canteen on the boat opened and the French bought it out. Canned salmon, tea biscuits, candies, orange pop, tomato juice. More than they'd seen in years.

A quiet trip that night brought us to Omaha Beach the next afternoon. It looked a little different. The Stars & Stripes waved from the hilltop. Roads were carved through the hill the 29th Division had assaulted for God knows how many hours back in June. We marched from the boat, loaded down with all we could carry. The first village was blown to bits—and the death smell of the Chateau was there. Three months later it waited for us. That was the first time I felt the old tingle in my legs.

For several days then, it was the Red Ball Highway. Jerry recon planes were being given flak while the drivers drove their 6X6 cargo trucks all night with lights on full, unafraid of Nazi aircraft. The storage dumps in the fields at this stage of the game were amazing. Two-acre fields completely covered with K-rations. In only an hour we marched to the road that made the newspapers scream victory when our troops reached it—two weeks after the invasion started.

When we got off the Channel boat, all the bed rolls were thrown down to the LCM in which we landed. One roll slipped and fell into the water. Oh, it was funny as hell. I laughed harder than anyone. It was retrieved with a hook and sent along. When our rolls caught up with us that night, mine was like a salted mackerel. Yep, the bag was mine.

Those truck rides all day in 6X6's were the wildest thing I've ever seen. Throwing K-rations to the French was the best sport. The rest of the time we huddled and tried to keep out the dust. The trucks were loaded to the seats with baggage and about 20 men rode each truck. It was a beauty.

Finally, after two nights of stopping at 10 pm, pitching tents and sleeping like logs until 6 am, when we were routed out, given lukewarm C-ration beans for breakfast, and sent on our way, we reached the 19th Replacement Depot at Etampes—20 miles south of Paris. Our Bn. Hqs. was set up in a great huge chateau. We inquired which one of the Louis's had once built it and were informed an anchovy millionaire from New York named Hyde had thrown it up in the 1800's. Ah, the glories of France!

Lamberson and I—still together—had our bags packed for an AWOL trip to Paris when the Bn. C.O. called us all in for a general chewing on the subject of AWOL trips to Paris. Seems a one-star general had picked up one of his lieutenants on same. Not in the least daunted, we were putting on the finishing touches for our trip after this chewing, when orders alerting us came out. That was that.

In Etampes I had run into George Bridgeman who was also on his way back to the regiment. He was hit D-Day or shortly thereafter, too. George had been sitting around an apartment he had wangled somehow for a few weeks and by now was an old hand on the Paris trip. Kelly and Flynn were at his camp and drank all day. They pulled their A uniforms out of a bag somewhere and became the snappiest combat joes in the ETO.

Dressed for the field, we boarded the 40 & 8 for Belgium, Kelly had bottles under both arms and one in his teeth. Some concrete light posts were emplaced along the railroad tracks about three feet from the sides of the cars as they passed. Kelly, in the flow of vin rouge, decided he was tougher than any concrete and kept trying to kick one of them by swinging his feet from the open side door of the box car. He finally made it. Broke the leather of his G.I. boot and his big toe in the bargain. Then started his sober reflection that if he reported it, they would accuse him of trying to be hospitalized rather than return to his outfit, the 1st Division. So, he hobbled and groaned until we left him.

The 40 & 8's are all they are said to be. We had ten enlisted men and three officers in our car. Found a slab of concrete on which we built a fire on the floor of the car. When we reached our destination, the heat had burned a two-foot hole in the floorboards of the train, but we were warm.

We got a look at Paris enroute in a tunnel. We came out long enough to throw K rations into apartment windows and be cheered when we whistled and yelled loud enough to attract attention. Nothing but heroes.

First main stop was Arlon, Belgium. Someone started firing a gun in the yards and we were alerted for kraut patrols from then on. The reconstruction job was amazing. Americans had been through this country only three or four weeks previously, and here we were in trains.

Finally, our destination in Huy, Belgium and Mud, Mud, and more Mud. Lamberson, Kelly, and I pitched tents together on a damp island and spent the first day learning how to burn green wood. Next day Lamberson and I got a ride to Liege. Went to the Town Mayor, gave him some fake orders, and were put up in a fine hotel. This was it! Ice cream. Beer with foam. Cognac. And flashes from the fight at Aachen at night.

We discovered the nightclub L'Observatoire about the first night. The proprietor brought all my French back with the aid of cognac, champagne, and orange drink. The MP's were on the alert. So was L'Observatoire. When the MP's started a raid, the doorman rang a buzzer. All Yanks headed for the cellar. The MP's looked upstairs. We kept quiet. It was a fine arrangement. We met the MP's there when they were off duty. They could get to the cellar quicker than anyone else.

The commandant of 9th Infantry Div. Hq. came in one night with a nurse—straight from Aachen. He tried to argue with the MP's instead of running. They ran him in. But he showed up again in a half hour minus all insignia. Without the eagles he looked as much a private as a colonel and had a helluva lot better time.

I was given honorary membership in the FFI one night after this had been going on for a few days. They checked the night clubs for ex-Gestapo agents but couldn't figure out how to approach American soldiers for identification. I got the job. We would check the place, then gather in the center of the dance floor for general handshaking, smiles, comparison of rifles, pistols and knives, and endless, "A bas, le Boche!" Vive le Belgique. Vive l'Americains. Also learned that the hip set always says, "Qu'est ce que c'n'est." Not "qu'est ce que c'est." Tres importante. ["What is it." Not "What is this." Very important.]

Lamberson and I became inmates of the Angleterre Hotel. The maids sneaked us into the royal suite for baths in one of the three bathtubs, and we generally had the run of the place. One night when we arrived happy and tired, the clerk remarked, "Pas les mademoiselles? Pff. Pff. Pas si bon. Pas si bon—y pas si mal. [No misses" Not so good not so good—not so bad."]

After four or five days of Liege, returning to camp only long enough to see if we were on orders for the following days, we were alerted.

About October 24, trucked up to a forward replacement battalion where it was really miserable. Rain and mud. Nothing else. Bridgeman went along too. He and I went up to the Regiment one day before our orders came out. Had lunch with the 22nd Regiment C.O., Col. Lanham; found Lum Edwards now a Lt. Col. and Regimental S-3. Saw Hal Simon, Claing, Newcomb and Twomey. No one else we knew. The faces had changed.

I went back to the 2nd Battalion and went to E. Co. for a day or two. Newcomb and Tolles just had their whiskey ration, and we drank it up in a dugout one night. We made a short move to Krinkelt and I was made S-2 on October 27th. Had the only house for a mile in all directions for Battalion CP. Rigged up electricity with the aid of some krauts and listened to a U.S. football game broadcast.

The 2nd Battalion Hq. staff now was: Col. Glenn Walker, C.O.; Tommy Harrison, S-3; Joe Samuels Exec.; Capt. Kerr S-1; Sisson, S-2; Andres,

Battalion Surgeon; Moffett, Motor Officer; Sgt. Maestro, Sgt. Maj.; Novak, 1st Sgt.; Edleburg, Scouts and Raiders leader; Evans, Intelligence Sgt.

At the 22nd Infantry Regiment Hq.: Col. Lanham, C.O.; Ruggles, Exec; Marshall, S-1; Blazzert, S-2; Edwards, S-3; Vance, Special Service; Rickerhauser, Liaison; DeHaviland, Field Artillery Liaison; Shugarth, I&R.

Jerry Claing had been fiddling with a rifle grenade and fired it with a live round in the chamber. Broke his leg nicely, and Phillips had F Co. now. Newcomb had E Co. with Tolles as Exec. Officer. A Lt. Mason in E Co. seemed familiar and suddenly I remembered he was center on that mighty football team we had at Camp Fannin when Southwestern beat hell, heaven, spirit and joy out of us.

The first couple of times I heard a 105 go off, I jumped a mile. Jerry was laying some artillery on us occasionally, and my intel officer job of sighting gun directions from shell bursts was great sport, of course, of course. I'd get my nose on the ground and sight a line, then get a tight feeling that the gunner on the same kraut 88 was getting ready to fire one more. Get an azimuth and get the hell out of there.

Got my first bath since Liege at a portable shower with Capt. Kerr. Things were sure a helluva lot different from the peninsula. At F Co. I found Pollock now 1st Sgt. and in pitiful shape, Montague, Morton and Baron in the kitchen—Kruzek, Scaming, Leichty, Darney and Sgt. Maines still around. Durham and Fitzgerald were at Regt. now. Got my first look at buzz bombs here. They were going about 200 feet over our house, headed for somewhere.

Cologne was to our front about ten or twenty miles. Saw a bombing there one night. Something I gladly missed. A red flare hung in the air as target marker, and huge flashes, like a giant photo flash, lit the sky at intervals for an hour at least. It was horrible. And beautiful.

Around the first of November, I was briefed on the coming operation, which was to take Duren via the Hurtgen Forest. Looked nice on paper. Several days later, I was on the advance party to the Hurtgen. Story was that Krauts were about a mile or less in front of us, and that we could make reconnaissance. Super-hush-hush. The 4th Div. men all wore 9th Div. patches. Jeep markings were all blacked out. Great stuff until one jeep washed the mud off the 4-22-1. We were known as the Ivy ghosts from then on. Hitler probably knew more about us than we did.

After a few nights in dugouts or pill boxes, the rest of the Bn. Hqs. came up and we put them into position. A beautiful night it was. Cold, snowy, sleet, freezing, foggy. Wait. Wait. Hours of waiting. Finally at dawn our Bn. came trudging along.

We set up in a pill box for Bn. command post. The 22nd Regt., with the 28th Div. on our right, were fighting for the village of Schmidt and getting a terrible mauling. Col. Walker sent me over to the 28th one day. I thought I had heard of artillery before. But when I saw patch after patch two or three hundred yards square with trees completely blown out, and sometimes a half mile with all the treetops blown off, I began to reconsider. At the 2nd Bn. CP of the 12th Regt., I found Col. Seibert who had just managed to get the remains of two of his companies out of a Jerry trap after they had been cut off for three days. I never saw a more beat up person. They were having it.

Finally, our attack order came. 1:10,000 maps. Minute planning. Made a reconnaissance one afternoon and Jerry missed us with mortars by about two hundred yards. Apparently, they were waiting for us.

At 12:45 on 16 Nov. we pushed off. Climbed the damndest hill I ever saw, then down again to tell Walker the assault companies were on top. No fight yet. The companies pushed on. The Battalion went up the hill and set up on top of the hill. Then we had it. Mortars for about five minutes. I looked at my hand and saw a little hole in my glove.

Brother, if there will ever be immunity to fire, Sisson won't have it. Patched it up and it didn't seem bad. We dug in on the reverse slope that night after firing and being fired at all afternoon. The krauts thought they were in this woods to stay. Every fifteen minutes, we got another barrage. Started getting casualties. Stopped to ask one soldier who was lying in a path if he was tired or hurt. He was neither. He was dead. Evacuation was terrible. The hill was a quarter of a mile long and at better than 45 degrees most of the way. Telephone lines were shot out as fast as they were put in.

The next morning, I made a reconnaissance to contact F Co. I got about a half mile and found one squad trying to dig in. When I asked them where the Company C.O. was, they pointed ahead. Then I said, "Where are the krauts?" A guy poked his carbine around a stump, sighted on something about a hundred yards away in a dense undergrowth and fired a few shots. "There," he said, "I think."

Oh, it was great in the Hurtgen.

The whole 1st Bn. staff was knocked out at one time that day. Walker finally sent me back to have the hand dressed. I would rather have gone forward than back into the range of those mortars again. It all ended up I had to go back to Division to get the hand repaired. And they evacuated me. I argued about being evacuated. That was once that I was really flak happy. Took off from the hospital the next day and got back to Division. That was the 18th.

On the way back, I learned the whole 2nd Bn. staff was knocked out after I left when I came to an ambulance control point. A driver said someone wanted to talk to me. I looked in the ambulance. There were Joe Samuels and George Kerr, walking wounded. They were kidding me about going back up. My bravery was melting fast. Then I asked where Col. Walker was. "Oh, he's over there in that other ambulance," Joe grinned, "he can't walk. Broken leg." That was that, and the 2nd Bn. staff was now reassembled. My confidence vice versa.

Got evacuated again shortly after them. This time to Liege. Oh, lovely day. By this time, I was all in favor of it. From then on, the war went my way.

After ten days I was released at Liege. Caught a ride back up to Zwiefall where the Division still was. In two weeks, the 22nd Infantry had advanced three miles in the Hurtgen Forest and had suffered 2,700 casualties. This was the same story for the 8th and 12th Infantries. When I got to Zwiefall I learned the 22nd had been pulled out two days earlier on December 4th and sent to a quiet sector in Luxembourg. On the same day, Col. Bryant, G-1, started pumping me about my records with the Division. He ended up saying, "Son, how would you like to spend Christmas at home?" Omigod, he meant it. For two wounds and ten days hospitalization each, a one each 30-day temporary duty in the U.S.A. was ordered by Eisenhower. I made it.

Followed then three of the worst days in my life, waiting for official orders to be cut. About Dec. 8th I get them. A daze, U.S.A., it said so. Signed, OFFICIAL: BY Command of...Zowie.

At Stavelot, Belgium where the 4th Infantry Division rear still was, we picked up more orders, left our weapons and got ready to leave the Division for Vervieres. This was the 9th. There I got my promotion to 1st looey. Things were rough all over.

Four days later, I understood Stavelot was no more. Something called the Bulge started and went through there like salts. Joy was going through me like salts at the time. I felt no pain.

The Hotel Edward 7, Avenue de l'Opera, Toneff, Gust, Marchand, Champs Elysees, Rand Pont, Marbeuf, Pigalle, La Vie En Rose, Claridge's, Cafe de la Paix, bicycle taxis; 700 francs from l'Etoile to l'Opera. Twelve days of it. Even missed the train at the end and got to fly across the channel instead of more 40 & 8's.

We landed through a pea soup fog and barely made it. The air corps, not expecting us, threw us off the field. No money, no orders, thirty men, three officers and Birmingham our destination. I started thanking the good lord for some experiences in the country before, pulled out my little book, got the 10th Replacement Depot Number, told them we were coming, got the men on the train and talked the conductor into giving us a lend-lease ride. Hell, these technicalities were a minor matter. If we hadn't been able to talk people into helping us, we would have had to make them, anyhow. Each one of us had a piece of paper which read, "U.S.A."

Several days at Lichfield again, and we caught

the train for Liverpool. Our ship was the Aquitania. That's all, brother. It takes seven days via Aquitania and a meandering course from Liverpool to New York.

With this, my father ended his wartime narrative, written during his leave in January 1945. He had been provided a precious opportunity to spend a month with his family in Ohio during some of the worst fighting during World War II in Europe. He would return to active service in Europe, but he would never again write about the war except in letters to his colleagues years later.

Part Five

IN MY FATHER'S FOOTSTEPS

"The great writer, Ernie Pyle, said that it is true that the majority of front-line infantrymen did not have the vaguest idea where they were during the fighting, nor did they really care—survival was the important thing!"[48]
— **James K. Cullen,** 2018, Staff Sergeant, Second Battalion, 36th Armored Division, WWII

Utah Beah, Normandy, July 2019[49]

21. My Journey Continues

2018

While editing my father's narrative, I found letters he wrote home to friends and family from his hospital bed in England, along with postwar correspondence with men who had served in the Second Battalion. I also found several books that I added to the ones I had initially identified. I still most frequently consulted the *History of the 22nd Infantry Regiment* compiled by Chaplain William Boice and Charles Wertenbaker's *Invasion*.

I decided to see what happened to F Company and the Second Battalion after my father was injured. I wanted to follow the men as they fought to seize the critical port at Cherbourg on France's Cotentin Peninsula. I knew that studying and mapping the battalion would continue to help me better understand my father's experience. My work documenting the Battle of Brandywine had included working with archivists and military archaeologists to identify and map the specific roads and cart paths used by the troops that descended upon the Brandywine Valley in 1777. This had brought the battle to life in a new way for people who used our maps when studying the battlefield landscape.

I developed a timeline of the key events in 1944, then used Google Maps to locate the towns where the 22nd Infantry Regiment fought, camped, or passed through as they advanced from Utah Beach to Cherbourg. Whenever possible, I identified the movements of the regiment's Second Battalion. I then asked a good friend and artist, Ann Bedrick, to render my map. Now I could visualize the troop movements through the end of June 1944.

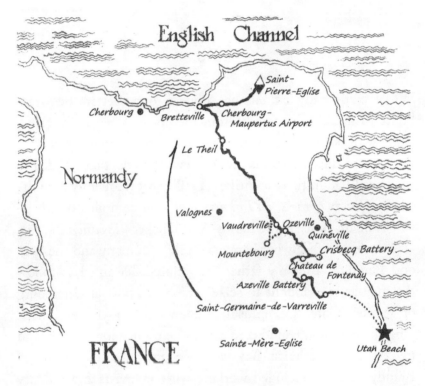

June 6–30, 1944: Utah Beach to Cherbourg

On June 11, 1944, the Germans abandoned their position at the Chateau de Fontenay. Chaplain Boice wrote that between June 12 and 14, the 22nd Infantry Regiment finally achieved its initial landing objective. Strong fire support, including naval bombardment, enabled the regiment to successfully assault and capture Ozeville, which was located on the high ground of the Quinéville ridge. The area was finally cleared on June 14.[50]

Charles Wertenbaker described the bravery of young lieutenants with less than a week of combat experience "walking calmly into rifle and machine gun fire," trying to persuade pockets of German soldiers to surrender and sitting in the middle of farm lanes studying maps while mortar shells burst around them.[51]

This action was part of the effort to secure the town of Montebourg, which was a communication center for many of the enemy forces on the eastern coast of the Normandy peninsula. Prior to mid-June, several attempts had been made to seize the town by other elements of the 4th Infantry Division, but the attacking units had been forced to withdraw before the town fell.

When the 22nd Infantry Regiment's Third Battalion followed the 8th Infantry Regiment along the railroad into the town around June 19, they were amazed the Germans had abandoned the town. Some three hundred French civilians came out of the cellars, having survived beneath completely demolished buildings during ten days of siege and terrific artillery and mortar bombardments. About thirty German soldiers, some in civilian clothes, and seven American soldiers were found sheltering in buildings. After a thorough search of the town, the Americans withdrew to higher, safer ground.

The words of a young officer, James Drennan, who served with the 42nd Field Artillery Battalion headquarters brought home the terrible price paid by civilians during war. He wrote:

> Walking into the city of Montebourg on or about June 14, 1944, I felt complete sadness imprinted on my mind. Dead mothers and fathers holding their dead children was a scene I was not ready to handle. Dead men from D-Day to the Battle of the Bulge have all faded from

my mind, but not the scene in the train or bus station. God, please forgive us.[52]

Montebourg, France June 1944

Wertenbaker astutely noted, "But the necessity for destruction did not make the destruction any easier to see, nor did it make it much easier to talk to Frenchmen who believed in us and had worked for us and were now shocked by what we had done."[53]

All through the summer, the wind off the English Channel was cold and dank. Supply replenishment was still sporadic, and the lack of an established front line made the fight even more perilous. My father was writing cheerful letters home during this time, but within his words the truth of what they were facing in combat conflicted with his desire not to further upset his parents.

June 20,
Dear Mother and Dad,

...Today is about the first time I have been able to really say anything and be sure it would get through. They've got some screwball security regulations which must be complied with. Was slightly wounded after several days and evacuated back here. We were going great and really had them on the run. Don't worry too much about our boys. We are getting a helluva lot more of them than they of us...The medics were having a swell time cutting clothes off the casualties over there in Normandy...The medics were doing a terrific job though. You simply can't imagine how well they come through...the number of boys they come right into the lines to carry out is amazing. Not that there are so many casualties—but the conditions are tough, to say the least...You can send as many packages as you want...

All my love, John

Between June 20 and June 27, the 4th Infantry Division, along with the 9th Infantry Division, would complete its advance north to seize the port at Cherbourg. For its part, the 22nd moved to the right of the 12th Infantry Regiment, bypassing Valognes and securing a strategic area near Vaudreville on June

20. From here they would move north to Le Theil to ready for the attack on Cherbourg. Chaplain Boice wrote that casualties were less, and resistance was crumbling. The 22nd would fight toward the town of Bretteville then head toward Saint-Pierre-Eglise to secure the Cherbourg-Maupertus Airport as part of the capture of Cherbourg. This was accomplished on June 25. Then, at last, the men had a respite from combat. Boice described the relief but also the cost:

> It was a good feeling to sleep without being shot at! The first hot meals since leaving England were served, and never had food tasted so good! Hot showers and the first change of clothing did much to raise morale. The first mail since the invasion was delivered, and the joy of word from home was dulled by the number of letters that had to be returned marked WIA or KIA.
>
> The fighting had been bitter and costly...two thousand and sixty casualties in less than one month of fighting, in a regiment whose normal strength was about three thousand two hundred men.[54]

I did not know how many casualties my father's Second Battalion had suffered within the regiment, but I saw how hollow a victory could feel. Wertenbaker further detailed the surrender,

> The men were sitting on the ground behind the wall in attitudes of utter weariness. They had been fighting and marching for two weeks with little rest.
>
> The Germans who held out at Cherbourg were a dwindling lot of stubborn men with dwindling resources of ammunition, terrain, and even medical supplies. But stubborn they were...Most of the men who fought under these leaders fought with Lugers at their backs. This was literally true, as captured orders proved. They were dock-

workers, Todt-workers, conscripts of subject nations, and plain Germans with no illusions that they were the master race or stomach for dying for it.⁵⁵

A very few rapists were hanged instantly and that stopped. Most of the [American] soldiers waved at pretty apple-cheeked girls in the streets and on country roads but behaved themselves well. The French began to like them. And so a delicate problem in international relations at last got off to a good start. ⁵⁶

After the fall of Cherbourg, the Allies would be increasingly exhausted, depleted, and overwhelmed as they tried to advance forward into a countryside that was defined by hedgerow-lined fields that were far more conducive to defense than attack. Outnumbered ten to one, the Germans fiercely defended their positions. Even the tanks that now accompanied the Allied forces were vulnerable and difficult to optimize.

Tank companies now reported that to advance 2,500 yards typically required seventeen tons of explosives to blow holes through nearly three dozen hedgerows, each defended like a citadel parapet. "There were snipers everywhere," Ernie Pyle reported, "in trees, in buildings, in piles of wreckage, in the grass. But mainly they were in the high, bushy hedgerows."⁵⁷

One survivor described it, "All the old values were gone, and if there was a world beyond this tangle of hedgerows...where one barrage could lay out half a company like a giant's club, you never expected to live to see it."

I said a silent prayer of thanks that my father had lived through his part in the Battle of Normandy. I thanked the good Lord and Al Mead. The soldiers' war was still too complex and brutal for me to grasp. Rape is terrible and unacceptable, but immediate hanging seemed equally so.

I kept thinking about James Drennan's prayer for forgiveness.

22. Return to Normandy

2018 – 2019

Battlefields and sites of armed conflict are powerful reminders of the shared heritage of all Americans...
       — American Battlefield Protection Program

During my deep dive into my father's wartime narrative and the 22nd Infantry Regiment's fight to Cherbourg, I had once again been struck by how limited my knowledge was of World War II. I was disappointed that I had not found any hints that could help me resolve the two soldiers' crossed claims concerning the *Mein Kampf* volume. But I had gained so much in the process. I had met my father as a young man and, through his descriptions in his story, gotten to know and like many of the men he served with. I had a deeper appreciation for the generation of men and women who lived through World War II, and I was much more aware of my father as a veteran, and the terrible price infantry soldiers and civilians paid during a war.

The final editing of my father's war narrative had coincided with the broadcast of Ken Burns and Lynn Novick's ten-part television documentary, *The Vietnam War*, in the fall of 2017. The fury of the war that had dominated my youth returned in a vivid rush as I was completing the edits to my father's story. Watching the familiar footage and endless national debates with my husband, I often felt so overwhelmed and conflicted I left the room. Who lied, who told the truth, who committed atrocities, who was right and who was wrong? The questions were left for the viewer to answer.

Memories of the titanic clashes at my family's dinner table revived again, this time as a reminder of how the national debate had raged in one home. Although I could not recreate one actual conversation from memory, I recalled the ever-simmering anger over not being "heard" and my moral outrage over my father's defense of the indefensible. My contempt was contagious. My brother and sister had turned to me as the family hero. They followed my lead, protesting against the war and authority in their own ways. That authority included the soldier with the two Purple Hearts.

When I stumbled onto a series of letters my father had written during the Iran Contra affair, I re-experienced the unassailable confidence he had in his beliefs that I had run headlong into as a 19-year-old grappling with my growing sense of conscience. In 1987, Oliver North, a highly decorated Vietnam veteran, became a polarizing national fixture as he testified before the U.S. Senate. The inquiry concerned his role from 1985 to 1987 in the secret arms sales to Iran to raise funds for the right-wing Contra guerrillas in Nicaragua, who were fighting the leftist Sandinista government. The U.S. Congress had banned government funding of the Contras.

On July 12, 1987, Dad wrote to Senator Daniel Inouye:

```
Lt. Col. North has done the nation a
great service by articulating—and I might
```

> say articulating better than any professional politician has—the real national interests involved. These interests include fidelity of purpose, continuity of foreign policy and the necessity for the U.S. to be perceived as a reliable ally. All these were jettisoned by a tricky end run for partisan purposes.[58]

As I considered my father's words, I decided to listen to him as a veteran, instead of focusing on our conflicting political positions or the illegality of North's actions. "Fidelity of purpose, a reliable ally." My research had made it clear that one of the critical priorities in war was to bring home as many soldiers as possible. To do this, plans must be well considered, munitions must be state of the art, and allies need to be effective and trustworthy. I could understand my father's position from this vantage point. I realized with a start that for the first time I had distinguished the soldier from the politician.

I thought about the angry years of my youth. One of my classmates died in Vietnam. One friend's brother came home uninjured, but we were told never to close a door in their house for fear it might slam. We did not witness his nightmares, but my friend did. I protested the war on both coasts and in between. I learned it's hard to set a car on fire and getting hit by a Billy club is a frightening experience. I remained undeterred in my opposition to the war and what I felt it represented. I did not question my actions or their impact on others.

My father worried the socialist element of the student protests would bring down our government. As far as I was concerned, he was part of the problem I sought to solve. This was the same father who had delighted me as a little girl after each business trip with an Oz book tucked into his suitcase, who had made the lion roar at the zoo, and who had bought us ice cream every Sunday after church. He had taught me to drive, bought me a car when I ran away from home, and protected me with no questions asked.

What had happened to us? To me? He was my hero when I was little and despite our differences, he was my moral compass for integrity as an adult. I loved and respected him. Carefully reading his story had helped me begin to bridge our differences, but I wondered if I could find a more concrete way to understand his worldview.

In the fall of 2018, Kathryn called me to say we had been invited to visit her mentor outside Nice, France the following summer. Thrilled, I extracted one promise. We must spend two days in Normandy tracing my father's footsteps on the narrow roads and through the hedgerows that he described after landing at Utah Beach on D-Day. I wanted to see and experience the landscapes and places within the military terrain that would have significance and meaning for an individual soldier—in this case, my dad.

After much deliberation, we agreed on an ambitious itinerary. We would travel the length of France in one week, allowing less than forty-eight hours in Normandy to rediscover 1944. We would retrace the route of my father's platoon, and we would return to the significant landmarks Dad photographed in 1979.

To prepare for the trip, I asked my daughter to read my edited version of her grandfather's World War II narrative, "*One Soldier's Story.*" I was struck and amused by her reaction: "It's kind of weird to meet your grandfather when he is your age. I think I got my sense of humor from him." And "I like Mead best."

Everyone likes Mead best, even Dad. Every war movie ever produced features Sergeant Al Mead. He walks into machine gun fire and battles snipers and even anti-tank guns one on one. The girls all love him, and he saves my father's life the night of D-Day. Dad described each man in his platoon with a paragraph in his story. Al Mead got two.

I had noticed that even in his narrative, my father had focused on the same stories of soldier antics that he told us about. Our visit would rely more on his 1979 return to Normandy, his pictures and correspondence with his old squad leader Bill Weingart, and several additional letters I had found. I used this information to map F Company's probable route from their Utah Beach landing to the Chateau de Fontenay where my father's Normandy story ended. The illustrated map Ann had created for me provided the larger context for the battle action.

The year passed quickly as our plans came together, and after a delightful visit in the south of France, my daughter and I conducted what I jokingly referred to as our harmless blitzkrieg of France. Map in hand, on July 3, 2019, Kathryn and I entered Normandy, seventy-five years after Lt. John Sisson and F Company ran ashore.

The cool air was welcome after the blasting heat we had experienced since arriving in France. In village squares we passed through, French, American, British, and Canadian flags snapped in the brisk breeze next to memorial statues and plaques honoring those who fought in both World Wars. Nothing appeared much changed from 1979, when the residents had met my father with warm respect. We saw interpretive signs that recognized and thanked the World War II soldiers who had liberated the region. Local pride in resisting the occupation and gratitude to the Allies for the pivotal role they played did not seem diminished by time.

Before leaving home, I had decided to visit the Chateau de Fontenay before we traced my father's footsteps from the beach. As we plunged into the narrow, hedgerow-lined roads, I began to worry. Chester County prides itself on its hedgerows that date from colonial days, but I was not prepared for Normandy's ancient gnarl of low trees and undergrowth. Veterans had noted in their stories that each section was its own fortress, but I had not understood how true that was. Sturdy stone farm complexes and walls provided the only relief from the unassailable bank of green in many places. How would I get my bearings? In June 1944, how did F Company?

I soon began calling out commands to my chauffeur daughter. "Not that turn—Pull over in the next village—This can't be right!—Turn around—Stop—Not here, go over there—I need to get a picture." Her favorite was, "No, we can't stop and buy gas." She was a very good sport. At one point, I wondered if my father issued his commands with so little regard for his platoon. I think not. Thank goodness I didn't need to shout things like, "Was anyone hit?—Krauts!!—What the hell just happened?" I felt at least as lost as my father in 1944.

Finally, we stopped at a narrow intersection close to our destination, just northeast of Azeville. Kathryn and I located an old farm entrance to the chateau which matched Dad's description of the path he and other wounded men took as they made their way back to the command post on June 10, 1944. We followed their route for about a mile to the next intersection, where a farmhouse seemed the likely aid station and command post for the Second

Battalion that day. The beautiful summer day somehow began to feel dark and foreboding.

How do you walk a mile with a bullet in your gut? What about the other wounded men?

At home, my map had seemed straightforward, but landmarks turned out to be hard to locate. Our GPS indicated the chateau could be reached somewhere close to us. It took several tries, but we found a narrow farm lane and drove slowly along the rutted path. We were deep into the grounds of the estate when the outline of the chateau at last began to emerge. I realized we were approaching from the rear. My father and his men had crossed the fields to our right on the morning of the battle. I took a deep breath.

Our journey ended at a T intersection behind the once formidable structure, now a vine covered ruin. Suddenly conscious of trespassing, we became aware of every sound. Unlike my father, we were not returning war veterans. A dog barking to our right called our attention to the proximity of several outbuildings. They must have been the ones my father said F Company ran toward as they prepared to attack on the other side of the chateau from where we were standing. Snipers were shooting at them from the structures we could barely see in the overgrowth to our left.

Kathryn and I got out of the car. For three days after the landing, my father and his squad leaders had led fifty men into enemy territory with no casualties. That all ended here on June 10, 1944.

The next day, the generals declared the beaches won and the land war begun. On that same day, only one third of the two hundred and forty soldiers who landed on D-Day with F Company remained on the field. My father survived a bullet to his gut and was evacuated. Many of the comrades he knew so well were not as fortunate.

J. Q. Lynd fought at the chateau with the 22nd Infantry Regiment and researched the history of the property and the battle. He had shared his findings with my father. Lynd had learned that forty-eight U.S. soldiers who died during the fighting near and at the Chateau de Fontenay were buried at the American St. Laurent Cemetery. Established by the First Army on June 8, 1944, it became the Normandy American Cemetery and Memorial in France. Another forty soldiers, originally interred there, were reburied at cemeteries in the United States. Poignantly, Lynd noted that, "The fate of some 14 others is unknown and remains within the listings of missing—known but to God."[59]

I could not help but reflect on the words of Col. John Ruggles, who served as Executive Officer of the 22nd Infantry Regiment in 1944, and eventually became a major general and the 1st Infantry Division's commander, after a tour as an advisor in Vietnam. "Sixty-four percent of all casualties suffered by American forces in World War II were in infantry regiments, which made up about 10 percent of the mobilized forces."[60]

We quietly turned the car around and left in silence.

After following the hedgerow-lined roads that crisscrossed the peninsula to our Air B&B, La Batterie du Holdy, we plunged once again into 1944. Our host had dedicated two rooms of the rambling old farm to the recreation of the nearby German artillery battery. Great care had gone into the authenticity of the rooms.

When Kathryn and I booked our rooms, I did not know the history of the battery. While researching our trip, I learned of the atrocities that occurred there during the night of June 5/6, 1944. Already shaken from the chateau visit, I began to question our decision to stay here, but I was beginning to understand that heroism and horror went hand in hand in Normandy. I reminded myself that I had come hoping to glimpse my father's realities across seventy-five years. They had made him someone I wanted to understand, and this was part of that reality.

Our host informed us he would be staging a reenactment of the D-Day landing at the battery that evening. Although it was sold out, he said we could experience some of the program from our room at the inn. It began close to midnight, so we headed out to dinner.

Entering Sainte-Marie-du-Mont, the first village liberated by the 4th Infantry Division after the D-Day landing, we paused before the great church that had played a strategic role. I read a description to Kathryn from *The Longest Day*.

> ...Pierre Caldron, the baker, saw paratroopers high in the steeple of the church waving a big orange identification panel. Within a few moments, a long line of men, marching in single file, came down the road. As the 4th Infantry Division passed through, Caldron lifted his little son high on his shoulders... Suddenly the baker found himself crying. A stocky U.S. soldier grinned at Caldron and shouted, *"Vive la France!"* Caldron smiled back, nodding his head. He could not trust himself to speak.[61]

Caldron was watching the 2nd Battalion of the 8th Infantry Regiment march by. I wondered if Dad's platoon with the 22nd Infantry received a similar welcome on D-Day when they marched through Saint-Germain-de-Varreville about eight kilometers north of where we were standing. He didn't mention anything in his story.

While enjoying dinner, Kathryn and I made new friends

when a table of residents realized I was having trouble ordering dinner in my halting French. They asked Kathryn and me to join them for dessert, and we happily accepted. They all were curious about the purpose of our trip and expressed a great affection for American war veterans. One of them, the owner of a store that featured military memorabilia, invited us to visit the next day. He encouraged me to write about my experiences and share them with others. I had not considered doing that.

After our return to the inn, the happy mood began to fade. Two German battery positions were located near Sainte-Marie-du-Mont during the occupation. The Brécourt Manor Battery has become famous because of *The Band of Brothers*. The Holdy Battery, where we were staying, was its heavily armed twin. Both had to be seized and destroyed to protect the infantry landing at Utah Beach. In the dark first hours of D-Day, several young paratroopers landed near the Holdy Battery. Some were shot and killed as they hung helplessly from the trees; others were taken prisoner and tortured. Their lifeless mutilated bodies were discovered by the soldiers who captured the battery at dawn and silenced the four 105 mm howitzers.[62]

At midnight, we experienced the reenactment of that cruel June 5/6 night in the woods surrounding the farmhouse. As we waited by our window, the sky suddenly lit up and we heard the roar of aircraft and bombs. Machine gun fire and explosions echoed through the yard and surrounding landscape. I tried to imagine how the earth shook from the massive explosions when the danger was real. This was overwhelming enough. We sat silently immersed in the battle until about 1:00 A.M., when things finally quieted down.

My daughter went to bed. I remained at the window. I felt Dad close by.

23. D-Day Revisited

JULY 2019

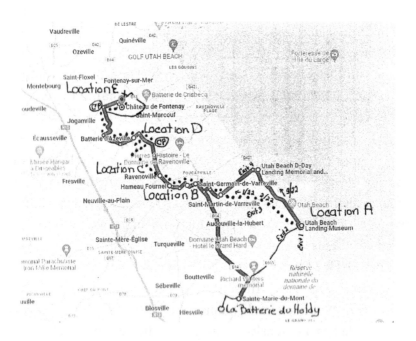

Kathryn and Karen Marshall Itinerary Map, July 2019

Kathryn and I awoke to a bright and windy morning. Chickens pecked at bugs near the entrance to the farmhouse where we were staying. We joined our hosts for breakfast and discovered that in Normandy what looked like cereal bowls were for dark, rich coffee. We learned that our host's devotion to honoring the liberation of France came from a deep sense of pride in his family's role in the resistance.

I was still recovering from the reenactment. I could not imagine what his family had endured and the risks they had taken. The terrible cries from the American soldiers being tortured and mutilated within the battery on the night of the landing echoed in my mind. It had happened so close to where we were sitting.

Does one ever forgive?

The cheerful room and delicious breakfast helped to dispel that dark question. We thanked our hosts and bid farewell. Our first destination was the Utah Beach access point at the Musée du Débarquement de Utah Beach (Utah Beach Landing Museum), which commemorates the Allied landing there. This is "Location A" on our itinerary map. I want to note that the map I have included is the one I completed at the end of our trip. Exploring Normandy was by no means a simple exercise in moving from one location to another. But in the end, often after several false starts and stops, Kathryn and I felt confident we had located our critical destinations and reasonable routes to get to them.

The D-Day landing at Utah Beach called for the use of four "exits" from the beach. These were dry roads that provided passage through the lowlands behind the beach that the Germans had flooded as a defensive measure. I indicated the locations as "Exit 1" through "Exit 4" on my map. The 22nd Infantry's battalions are numbered 1/22, 2/22 and 3/22, the arrows show the direction each moved after landing near Exit 2. When my father came ashore with the Second Battalion's F Company, he remembered that the road at this exit was already jammed with men and vehicles. The dots on the map represent my father's footsteps as closely as I could approximate them. The solid black lines are roads Kathryn and I followed.

From the museum parking lot, Kathryn and I headed into the brisk wind toward the dunes and the famous beach beyond them. We stopped before a statue of a landing craft whose soldiers were running toward the unknown, toward the enemy, toward Occupied Europe. Overhead the flags of the nations that had fought here snapped in the wind.

We continued to the beach, then stopped so I could read an excerpt from my father's narrative. I would do this often over the next six hours of discovery. I wanted to hear his voice in the places he had written about.

Although they would face terrible fighting in the days and months ahead, the morning of June 6, 1944, would be a decisive victory for the 4th Infantry Division on Utah Beach. From here they would begin to pour inland to free Europe. I read my father's words aloud:

> ...On the night of June 5th, we were out to sea with England out of sight. We glimpsed the other convoys from time to time, and the channel was literally afloat with ships...We could see Omaha off to our left—our own Utah beach dead ahead. Battleships were sending up great flashes of smoke and flame as they threw shells shoreward. Rocket boats disappeared in a cloud of smoke as they discharged their deck full of rocket launchers. The sky was full of buzzing planes, and we could see them dive down to strafe the

> beach…But there was an air of unreality
> about it.

Kathryn and I walked to the water's edge.

We looked down the beach and saw a house in the far distance. Was this the landmark that confirmed to General Roosevelt that they had landed at the wrong place? "At 7 A.M., he informed the 8th Infantry Commander, Col. James A. Van Fleet, 'You see that brick building over there to our right front? It always showed up in those aerial photographs, and it was always on the left…I'm sure we're about a mile or two miles farther south.'"[63]

What he didn't say was that they were also significantly south of the defensive batteries the 4th Infantry Division was to subdue. As a result, the German artillery could not reach the men landing on the beach with the same dire consequences as was happening at Omaha Beach. Standing on the long, flat beach devoid of any protective features, I shuddered to think of what would have happened if the division had landed in the correct place or if the airborne forces hadn't subdued most of the smaller artillery batteries the night before.

I also gave a prayer of thanks to the men of the 8th Infantry Regiment who landed at H-Hour (6:30 A.M.) with the 22nd Infantry Regiment's Third Battalion. Together with Army combat engineers, light equipment, tank battalions, and Army and Navy demolition teams, they cleared the beach. Their work to neutralize defensive land mines and fortifications and their grenades, tommy guns, and tank fire resulted in my father's straightforward description of his landing around 9:30 A.M. He had noted a few artillery shells landing close by, but the men were able to move quickly from the beach.

Kathryn and I lingered. We tried to imagine thousands upon thousands of men, landing craft and equipment on the now empty beach. German shells would be exploding close by, vehicles getting stuck, uniforms and supplies wet, and everything laden and gritty with sand. It was much easier to picture Captain Fulton and the two hundred forty soldiers in F Company amid all the confusion. My father was first off the beach with Captain Fulton. His Third Platoon of almost fifty soldiers led the 900 riflemen of the Second Battalion. Sergeants Weingart, Wirtzberger, and Durham were his squad leaders. Sergeant Mead was the platoon guide. Directly under my father was Jim Conly, Tech Sergeant. I felt a deep pride for all of them.

We were about to continue our journey with F Company's Third Platoon when a guide we stopped to speak with encouraged us to visit Le Roosevelt. This simple fisherman's house survived the landing and is now a bar and café that also memorializes the role it played for the Allies as a communications center for the Navy. The men who staffed the center signed their names on the walls in 1944. Over the years, hundreds if not thousands of D-Day veterans have added their own signatures. After describing my trip to the bartender, he invited me to sign on behalf of my father. It was an honor I will never forget.

F Company did not follow a road as it left the beach. Maps published by the U.S. Army's Center of Military History show that the First and Second Battalions turned northwest from Exit 2 and marched through the inundated area toward Exit 4, then found dry ground, most likely on highway D-14, which leads to Saint-Germain-de-Varreville. Because we could not follow on foot, we stood on the beach at Exits 3 and 4, then drove around to rejoin their route in town (Location B on the itinerary map). It seemed odd that my father had not mentioned the inundated areas, with their miles of swamps and water in some places up to the waist—over the head in irrigation ditches.

As Kathryn and I entered Saint-Germain-de-Varreville, it was hard to believe we were actually following F Company. I could picture the exhausted paratroopers my father described and the tension and the confusion. Did the citizens hide or greet them? From here, Captain Fulton, Dad, and some of the soldiers from his platoon set out into Normandy as the advance patrol for the Second Battalion. I had determined their objective was to rendezvous near Caufour, where the Second Battalion (minus Dad's patrol) was to set up a defensive position for their first night on shore.

> We posted four guides behind us to direct the remainder of the battalion—and of F Company—and struck off down the road with about fifteen or twenty men, including George Caldwell who was our artillery forward observer. We marched steadily for about two hours and reached our rendezvous, but had completely lost contact with any other troops...

I wondered how Captain Fulton and my father had selected the fifteen or so men for the patrol out of the two hundred who made up the balance of F Company. I guess it really was "F Company against the Wehrmacht," with the aggressive support of the paratroopers on that first day. At least this was what my father had described in a letter to Dick Taylor from his hospital bed in England. That letter had helped me clarify some of the brief descriptions in his manuscript. Its bravado and profanities seemed much more fitting than the language he reserved for letters home, and even his narrative.

He told his old school friend about the amazing good fortune of landing at the wrong place on the beach and totally surprising the enemy. He noted they saw so few Germans that "for a while we thought we had invaded England."

> We hit about the tenth wave, I guess. Each bunch of crafts to beach is a wave. I negotiated the beach soaking up a ton or so of sand. Our beach was a lulu, and the Krauts had all scurried off, having seen the die cast.
>
> We were all screwed up alright. But they were so screwed up that they kept running around in ever decreasing circles. I'll tell you one thing. NEVER, NEVER, NEVER, mess with any paratroopers!
>
> Incidentally, in case I didn't tell you, you can forget about half the things I told you to bring over. They had us loaded down to the hilt when we landed but within a couple hours everyone was rid of everything but a belt, ammunition, a couple boxes of rations, some socks in their helmet, and maybe a parka and a raincoat.[64]

I had smiled thinking about how many millions of dollars' worth of equipment lay piled up on and beyond the landing beaches.

We felt pretty certain that the narrow farm lanes and dense hedgerows near Hameau de Fourneau (between Locations B and C) were where Captain Fulton left to contact the rest of his company and Caldwell decided to shoot the first round for his platoon. Dad sounded so calm about standing in dense foliage with a small group of armed soldiers while snipers fired all around them. I thought about his entry when he described how the men had absolute faith in General Eisenhower. I wondered if it was this faith that kept them so purposeful.

Our next stop, Location C on the itinerary map, was the site of the platoon's most successful action under my father's command. I had located its likely spot thanks to a letter written in 1989 to Lt. Gen Glenn Walker, who was a captain when he landed with the 22nd at Utah Beach on D-Day.

> You said you were going to Normandy and that you didn't have any maps. If you go to Normandy without a map, you're just a tourist. With a map, perhaps you can try to run a company or a battalion again…On D-Day night my platoon found itself alone along a road where we had a chance to shoot up some Kraut vehicles and an 88 being towed. Looking back on it, I think we were on the Ravenoville—St. Mere Eglise Road, although we didn't have the foggiest idea where we were, having gone across country all afternoon.[65]

It looked like he was referring to D-15, and the most likely intersection was with D-420. Once we got to the intersection and had parked the car in a farm path, Kathryn and I walked back to our destination and read:

> ...we heard the popping of some sort of vehicle coming down the road. We hardly got a look at it when Servat let go with his BAR at it and things began to happen in a hurry...Something exploded, and I figured Jerry had some artillery on us. Weingart and I looked at each other and both reached for a grenade at the same time. At that precise second, all hell broke loose around there, and shells, shrapnel, and fire started shooting in all directions. "Weingart," I said, "let's get the hell out of here." We hit a shallow ditch about fifteen yards away as one man.
>
> As it turned out, we had hit an ammunition truck, and it went along merrily for about fifteen minutes until another vehicle came along. Servat got this one too. We let the fire burn out and rounded the remaining Jerries up.

In another letter to Dick Taylor, Dad summarized his satisfaction with the reaction of his platoon to their first action:

> I always figured in training it takes no dry runs to learn to dive headfirst through a briar bush into a ditch almost 25 yards away in .00000127 seconds. (In

> this period of time I noted that even the
> laziest man in my platoon who had spent
> four years to get laziness down to a sci-
> ence was long ago under cover.)[66]

He noted that when they reviewed the situation after about a half hour of explosions, the platoon, "drew first blood" when a German soldier tried to run away. This was also when Sgt. Al Mead saved Dad's life.

> My carbine jammed just as a Kraut jumped
> up from the ditch about four feet from me
> and swung his rifle around to fire. He fired
> once, but not again. Mead was around as
> usual and took good care of him. We or-
> ganized the company around the crossroad
> and dug in for the night.

Thus ended D-Day for some forty American soldiers from the Third Platoon, F Company, 22nd Infantry. It seemed almost incomprehensible that after the twenty-four hours he had just witnessed, my father could end his narrative so simply. But how do you describe what it feels like to watch the young man who was going to kill you be killed?

After all the times I had read this part of the narrative, I wondered why I had not considered this before.

24. D+1 and D+2

JULY 2019

The morning of June 7 or D+1, the Third Platoon assessed the five disabled vehicles, including the fearsome German 88, which completely blocked the road. Reading my father's words, I could sense an unstated pride in their accomplishment. After rejoining F Company, they began to send out patrols and experienced increased enemy action. Lt. Jim Beam reported watching from a hedgerow as a five-man German patrol marched by him. Prisoners were taken; Dad reported that one, Smitty, carried the radio for the afternoon.

> It was wide open going most of the time, with no fixed lines, snipers all around—those goddamn screaming meemies that are worse than a whole regiment until you get used to them, if ever. That was the sort of fighting it was. No one knew where anyone else was—friendly or enemy.

Later that morning, Captain Fulton ordered an advance toward Azeville to the west. Kathryn and I could only approximate the route they took between Locations C and D on our map. We explored the narrow roads; it was beautiful countryside with a dark secret. At one point we stopped, and I read descriptions of war in the hedgerows to Kathryn:

> Henceforth the war would be fought inland, through the biting dust of dry roads and the clinging mud when the roads were wet, along shallow ditches beside the roads, from hedgerow to hedgerow across mined fields, up and down rolling hills, into towns blasted to rubble and over rubble from wall to wall, from foxholes and slit trenches and bomb craters. [67]

It must have been late afternoon when Captain Fulton sent my father with a patrol to find the Second Battalion. Ready for anything, the patrol did not meet serious trouble and found the battalion. My father recalled,

```
They didn't know where we were, and
Maj. Edwards was rapidly getting jerk-
ed off. We pulled into the battalion left
flank, and my platoon set up flank secu-
rity around a farmhouse. We got our first
crack at fresh milk since we had left
the States and some of the first eggs in a
long time. The farmer tried to marry his
daughter off to Mead, I think. He figured
he was quite a lover.
```

What my father did not know was that while F Company was scouting on June 7, the First and Second Battalions had been ordered to attack the Crisbecq and Azeville Batteries.

Our two battalions now faced the two most powerful coastal forts… Each position consisted of four massive concrete blockhouses in a line; they were supplied with underground ammunition storage dumps, interconnected by communication trenches, and protected by automatic weapons and wire. An arc of concrete sniper pillboxes outposted the southern approaches to Azeville. [68]

The attack on the Azeville Battery by the Second Battalion began early in the morning. Chaplain Boice described the situation.

> The next morning, June 7th came and despite the falling rain, the troops slept. They had fought doggedly until very late that initial day. At the first light of dawn, it was possible to see the non-coms and officers organizing their units for the oncoming attack. There was a tenseness in the air, for the men had lost the thrill of battle hours before…In the distance could be heard the tat-tat-tat of machine guns and the echoing roar of the artillery pieces.
>
> The enemy was a stubborn and determined defender. From every conceivable defensive piece of terrain were German guns, stirring the cold early dawn with their terrifying crack. Overhead, allied planes continued to roar inland…[69]

The attack was violently rebuffed, and the battalion suffered serious casualties. Boice recounted it was "suicide" to continue the advance. They fell back to the farmhouse where my father reported joining them. I directed Kathryn to a large livestock breeding farm on D-420 about a mile southeast of the Azeville Battery. This was where I believed the Second Battalion and its headquarters were that evening.

As night fell, the Germans staged a counterattack:

> There were German automatic weapons firing
> everywhere—and things were looking bad
> when our mortars suddenly opened-up. They
> walked back and forth across our front
> and then began to chase the Jerries back
> to their holes.

On June 8, D+2, F Company was part of the second attack on the Azeville Battery. For this part of our journey, I relied on my father's account of his visit to Normandy in 1979 in his letter to Bill Weingart. My dad had described Weingart as a kid of twenty or so from Brunswick, NJ who was far and away his best squad leader.

Kathryn and I had little trouble finding the hedgerow and hill my father referred to and photographed in his letter, Location D on our itinerary. We stood, as he did in 1944, and again in 1979, on the unnamed farm lane that runs parallel and south of D-269 below Azeville, looking up at the field and hedgerows leading to the first fortification. It was a structure so powerfully built that when the battleship *Lexington* made a direct hit "the paint wasn't even scratched," according to my father. The wide-open fields that led to the mammoth fortification only made me more astonished at the bravery of the soldiers.

Standing silently in the field, I reflected upon the planned attack on the Azeville Battery by the Second Battalion. Where had these largely untried young men found the courage and conviction to assault an entrenched enemy in an occupied continent. Two days after the landing, they still lacked necessary support. Chaplain Boice's description of the battle drove home their perilous situation. He wrote that as casualties mounted, there were no replacements to take their place. "As a result, hitherto unknown leaders suddenly sprang forward, privates were commanding squads and sergeants were leading platoons."[70]

Sergeant Mead was coming to embody this rugged spirit for me. Kathryn read to me from her grandfather's story:

> The next day we pushed up directly under Azeville, where there were four casemated 155 rifles in battery. I got orders to go up a wide-open hill with my platoon and take the battery. An order to pull back and prepare for a counter-attack postponed that one. It took a battalion to take it the next day—and they approached from an entirely different route. Mead was pulling his usual stuff. He took the sniper's rifle away from our sniper and started a personal battle, at 200 yards, with a 20 mm or 40 mm anti-aircraft gun which was in position on top of one of the pillboxes.

In his 1980 letter to Bill Weingart, my father explained the pictures he had taken of the Azeville Battery and provided details of the day. When I had read Dad's narrative, I suspected that this was one of the moments after the landing when my existence hinged on total chance, but I wasn't sure. The open field and minimal tree line were visible in his pictures at home, but I had no idea what the fortification looked like—it was hidden in the trees.

Standing where my father had fought, looking at his pictures and reading the description of the attack, I no longer had to wonder. Kathryn and I had toured the Azeville Battery before we traveled to this quiet hillside. If the order to continue up the "wide-open hill" had not been rescinded, it was quite clear the Third Platoon would not have survived. Without any cover, the men were to charge guns that blew holes through tanks and ships. I could not fathom how anyone in the platoon summoned the courage to move forward.

To convey what we saw; I have juxtaposed the picture my father took in 1979 from where the attack began—the battery is to the right of the tall tree—with my picture of my daughter standing in front of the same structure. Even at six feet tall, she is dwarfed by its size. This was just one of four installations in the complex.

John Sisson's Picture #1, 1979

Kathryn Marshall, 2019

My father wrote to Bill: Picture #1 is a shot of the pillbox you started after which was taken from the road along which we approached. If you recall, your squad started up the hedgerow which shows on the left of Picture #1. Al Mead was along the road somewhere to the left in the picture, firing the sniper rifle at a Kraut on top of the pillbox who was trying to swing a small anti-aircraft gun around on us. I remember checking with you along that hedgerow when you had a

```
cartridge jammed in the charger of your
rifle.
   I could see the second bunker to the
right of the one shown, there were two
more across the road. After visiting the
scene, I could see why someone wised up
and pulled us back, allowing the 3rd Bn
to take the bunkers by coming in on their
unprotected left flank.⁷¹
```

Bill Weingart responded: Picture #1 seems about correct but there is a lot of foliage there now that was not present then; the Krauts had an unobstructed field of fire back then. We were making a frontal attack on those damn things and the only thing that saved my squad was the fact that the German on the 88mm could not traverse right any further, but he sure was plowing up that field 30 or 40 yards to my right with direct fire.⁷²

Gazing up the hill toward the battery, I couldn't help wondering if the command that had ordered the attack had any idea of what the field of battle looked like, or just how well defended the battery was. I remembered a statement I encountered during my research concerning the sharp contrast between the intense planning that went into the D-Day landing versus the paucity of plans on what to do after the troops advanced inland. This seemed a good example.

On the evening of June 8, the decision was made to focus the 22nd's three battalions on the Azeville Battery. The Third Battalion was given the order to attack the next day and received full credit for the fall of the battery on June 9, D+3.

In the end, it was one private's resourcefulness, bravery, and luck that defeated the fortress, not headquarters' plans. After the rear recessed doors of the battery resisted three direct flamethrower attacks, Pvt. Ralph G. Riley was sent with the last flamethrower to the blockhouse:

> Pvt. Riley ran seventy-five yards under fire and dropped into a shell hole for cover. The flame thrower would not work, and he tried to think of the proper "immediate action." He opened the valve, held a lighted match to the nozzle, and trained the stream of fire on the base of the door.[73]

Just as enemy artillery fire from the nearby Crisbecq Battery began and all hope looked dim, Private Riley heard a popping sound and then explosions. His stream of fire had set off the enemy's ammunition within the blockhouse and soon a white flag was raised. The German commander surrendered all four forts and their garrison of one hundred sixty-nine men.

I felt that if they had let F Company have a go at the back door instead of the front, Mead might have accomplished the same. Mead and Riley were cut from the same cloth. I'm sure it was no accident Private Riley's company commander selected him for the last-ditch mission. I hoped this brave young man returned home safely. (Editor's note: He did, I have a copy of the small book he wrote.)

At this stage in the war, 4th Infantry Division Headquarters was unhappy with the progress the 22nd Infantry Regiment was making. Their commander, Col. Tribolet, was relieved of command on June 10. The men disagreed. They felt they had given their all under impossible conditions. In his history of the regiment, Chaplain Boice made a special point of noting, "'Trib' had trained them; he knew many of their families, and he was both loved and respected by the officers and men. Their regard for him and their confidence in his ability and leadership remained."[74]

I agreed with the men. I was beginning to understand the difference between the generals' and the soldiers' war.

25. One Battlefield in Normandy

JULY 2019

The Second Battalion moved northwest from Azeville on June 9, 1944, to prepare for an attack on the Chateau de Fontenay on the morning of June 10, D+4. This was part of the reduction of the Azeville area fortifications. It was early afternoon when Kathryn and I prepared to follow F Company's footsteps from Azeville to the chateau on the route my father had mapped in 1979. I was relieved we had gone to the chateau on our first night. I felt overwhelmed by all we had seen and experienced and did not want to be struggling with directions. (Location E on our itinerary map).

As we navigated the narrow country lanes with their unforgiving hedgerows on either side, I remembered my father's early training in England: Fields are safer than roads. The hedgerows were impenetrable in many places; there would have been no escape from an enemy attack. Our destination was the intersection we had found on our first night. This was where Dad had placed a flag on his map to indicate the location of the command post in 1944. He had circled the chateau and the Azeville Battery south of it. I was almost certain that the X beside the circled chateau was where he was injured during the battle.

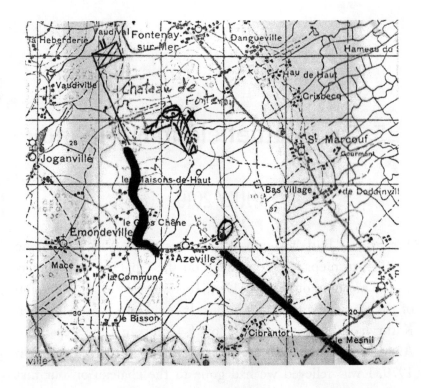

John Sisson's Map, 1979

Kathryn and I returned to the small stone farm complex I had identified as the likely site of the battalion's command post during our first day in Normandy. No longer lost, I knew we were at the edge of the field through which the men had approached the chateau. Looking at the aerial view of the chateau from my father's files, the challenge faced by the Second Battalion became clearer. The Germans were concealed in the woods with an unobstructed view of their approach. The chateau and outbuildings offered some cover, but there were snipers in the trees. The overgrowth we saw was not there in 1944.

The Americans advanced along the arrow towards the chateau, which is almost in the center of the picture. The Germans were in the tree line marked by the slashes. The Second Battalion, led by F and G Companies, advanced toward and then around the outbuildings of the chateau, then crossed the open lawn directly in front of it to assault the German position. Casualties were very heavy, one platoon lost all but two men. The X is where my father was injured, near his platoon's farthest advance point. This was our next destination.

Aerial Photograph unknown origin, John Sisson files.

After driving around the estate, we turned off the road and entered a great canopy of trees that sheltered the main drive leading into the property. It was a dark and quiet world. Our voices dropped to whispers; the car lurched from side to side as it struck ruts in the road. I was anxiously looking for someone who could give us permission to visit.

Chateau de Fontenay, John Sisson photograph, 1979

And then we were there, where my father fought in 1944 and returned to in remembrance in 1979. The chateau was now covered by foliage, but the view was otherwise the same as the picture he took in 1979. We got out of the car. I read aloud from my father's letter to Bill Weingart, in which Dad described the battle on June 10, 1944.

```
This is a view looking back toward the
Chateau from our position after cross-
```

> ing the grassy area. You can see the…moat which is marked by the stone vase monuments. It shows the outbuildings behind the Chateau where we first entered.

In his history, Chaplain Boice recounted that Lieutenant Beam had warned Captain Fulton not to climb up on a pump house, one of the outbuildings beside the chateau. The incoming machine gun fire killed the captain just as he "dropped down with a smile and his usual ready retort."[75] It was a heartbreaking description. Looking at the ruined buildings, I wondered if this was where that charming young man was cut down—never to see the wife he loved so much again.

The letter continued:

> We pulled around to the left and rushed across the grassy area. As I understand, you were pulled back later that day for a bombing of the Chateau. The flyboys did quite a job, as you can see. The irony may be that the Chateau is surrounded by a water moat, we couldn't get up to it, as I recall. We didn't get any fire directly from the Chateau as we came across the grassy area. The Krauts seemed to be behind the Chateau and to the right in the picture.

I knew from the pictures I had left at home that my father had taken this picture from the intersection of two dirt roads on the edge of the wolods where we were standing. Facing the chateau, the German position was to our right in the woods. Just behind us was the ivy-covered stone wall and pillar, the high-water mark of the attack.

> This somber spot brought back vividly that Ozzie Wirtzberger's squad had taken a German anti-tank gun with its crew, just behind the ivy-covered pillar. Ozzie was lying in the grass just in front of the wall. Ralph Carter was wounded, lying across the wall under the trees. Jenkins and I dressed his wound at the chateau and Jenkins stayed by him for over an hour. We couldn't get litter bearers up for either him or Wirtzberger.
>   I recall trying to get Lt. Camper to set up a light machine gun at the corner of the wall. The Krauts were dropping mortar and light arms fire into the trees. I was hit while trying to move up along the edge of trees and the open field behind the pillar.[76]

In a 1989 letter to Lt. Gen. Glenn Walker, my father provided a few more details about the confusion and lack of support his platoon experienced:

> You asked if I knew what happened there, where Capt. Fulton and Lt. Cook and others were killed. We got past the chateau; the Krauts were not there in any force;

we captured an anti-tank gun at the final
line of our advance, where I was hit. We
contacted no one else—G Co. was supposed
to be on our right. We didn't see or hear
from them. The radio wasn't working—we
couldn't get mortars or light M.G.'s.
I walked back along the road to the
CP—didn't see anyone enroute. Contacted
Lum at the CP to report. He said the road
I walked back to the CP had been under
heavy fire. I told him it wasn't now. We
had heard some firing when we were going
around the chateau, but it didn't seem
that much. I think it is likely Fulton
may have been along the road somewhere,
although neither I nor a couple of wounded with me contacted anyone when we went
along the road.[77]

From his description, I realized that Kathryn and I were standing in the lane that Dad and other wounded men had used to walk back to the command post. We could still see the outbuildings my father said his Third Platoon cautiously came around before they rushed across the open grass toward the German position in the trees on the far side of the chateau. They had to summon the courage to run straight into fire from a well-hidden enemy. I winced thinking about how far they had to run without cover.

I could almost hear the mortars and light arms fire my father described. It felt haunting to stand before the ivy-covered pillar where Sgt. Ozzie Wirtzberger's squad took the German anti-tank gun with its crew and where Ozzie and Carter were killed. Dad didn't know where Lt. Cook, "Cookie," lost his life during the battle, but he did note Sergeant Mead was injured close by:

> Mead was hit at the chateau too. A machine gun was spraying the side of a building near a gate—but he thought he might be able to get a shot at it—so he walked right into the burst and caught one in the hip.

I was deeply affected by what we were seeing and reading, something I had not expected. When I had studied the movements of soldiers in the past, they didn't have names, let alone any relationship to me. Of course, poignant scenes in war movies elicited strong feelings, but this was different.

A vague memory began to nag at the fringe of my consciousness. I felt like I was missing something.

Lt. Franklin Shaw joined F Company with the Second Battalion "just after the disaster of the Chateau de Fontenay."-He confirmed that Harold Fulton, the company commander, and one lieutenant had been killed. Two other lieutenants had been wounded and evacuated and some ninety men were lost. Shaw noted that Lt. Jim Beam was the new company commander. Shaw may have replaced my father. He was wounded just four days later.[78]

In response to my father's letter, Bill Weingart had provided the last details of the battle I could find:

> Sgt. Cobb and his First Platoon got beat up pretty bad there. Edwards

was killed I believe and Lightcap wounded or killed, I am not sure. Deegan was captured along with some others. I lost Lester Komowski, my BAR man, and of course Wirtzberger was killed. I crossed that grassy area a couple of times with Beam back and forth, back and forth, and finally we got something going in the right direction.

That night Beam gave me Cobb's platoon or what was left of it; Challis Peton, Norm Stanley and I took Paul Anish from your platoon, and they gave me a platoon of replacements. Days and time are all foggy now; I know I got hit on the 23rd of June up the coast nearing Cherbourg and never went back…Beam and I enjoyed the two small bottles of Scotch you had in your pack; I think I told you that.[79]

My daughter whispered she wanted to leave. She had started crying in deep trembling sobs. It shocked me back to the present. I had been wondering if Dad had brought those two bottles of scotch all the way from England. That was so like the man I loved and missed. Although he never got to savor them, it seemed fitting that Weingart and Beam did, recovering whatever sense of balance they could.

I hugged Kathryn tightly and smoothed her hair. Walking back to the car, I looked at the fields one last time and wondered which were the hedgerows Chaplain Boice had mentioned when he wrote:

> At the close of day, the remnant of Company F was re-grouped in the hedgerows before the Chateau. Only a third of the company remained under the command of Lt. Beam. Casualties had been extremely heavy, and throughout the regiment, men were face to face with the fact of violent death for many of the men they had trained with for three years. It was a shattering emotional experience that hardened the attitude of the men toward such an enemy.[80]

We turned the car around and started to leave, but I asked Kathryn to pause as we passed the ivy-covered pillar. Ozzie and Carter had fought to the farthest point of the attack for the Third Platoon of F Company and seized a German anti-tank gun here before they were killed. I could see the field just beyond where my father had been critically wounded. And there were so many other young men killed or wounded. I felt an ache in my soul so deep that it seemed hard to breathe. This was heroism, and this was its cost. I was in awe of their bravery.

Suddenly the nagging memory broke through the silence in the woods. *How could you understand when you don't care! We are killing people who don't want this war. The military is a machine, it only answers to corporate America and corrupt politicians! This is an illegal war, forced on the public by people like you who ignore the human cost!*

It was 1969. The war in Vietnam was raging. I was standing in full fury beside my father at the dinner table. I saw, as if for the first time, that his head was slightly bent, and he was not looking at me. I had always assumed that his silence sometimes after an argument reflected his indifference to the matter of war. I was usually storming from the room and not paying much attention. A deep shame swept over me. The machine I was referring to was filled with living and breathing men my father had known and cared for. As for the killing, he understood that all too well.

My daughter looked at me with a troubled frown, "Mom?" I shook my head and tried to smile. She started the car and we left. I didn't dare speak. My heart was pounding. I now understood why Dad was only willing to talk to us about his antics in the army and why seven months after the battle all he wrote in his narrative was,

> On the fifth or sixth day we attacked Chateau Fontenay or some name similar to that. Jerry got on our flank and a sniper crept in to within thirty yards of me and got me too easily. This started a long trip back through channels.

This was despite the fact that thirty-five years later he could remember most of the details when he wrote to his squad sergeant.

I would never again wonder why his narrative paid tribute to the medics who saved his life and the men of his platoon instead of recounting combat details. This is what he said about the men who died at the Chateau:

> Ozzie Wirtzberger was an easy going dutchman with a wonderful sense of humor. He was from Frankfurt, Kentucky and loved to recall the better and bigger beers of the Ohio River Valley. Even the invasion had a hard time breaking up the pinochle game he presided over. At the Chateau, he led the attack. A sniper caught him across the top of the head. He's buried near St. Mere Eglise, with Carter who was hit twenty feet from him.
> Ralph Carter was a Pvt. in the rear rank. But he was morale. If any statistics on the Detroit Tigers or University of Michigan's football team escaped him, they failed to exist. Anything was worth

arguing about, and his lack of information or logic on the subject never bothered his volume. He mowed 'em down. About five foot four, he was a perfect likeness of the Stars & Stripes Hubert.

Capt. Fulton and Cookie live in France now.

Captain Fulton was just married and wouldn't take a pass, because he could think about his wife just as well in his quarters as wandering out around the bright lights of the big city.

When it came to girls, Cookie was always the pappy to the company and had his usual following.

My father had closed his letter to Bill with his thoughts on how to honor and remember those who lost their lives:

It would be most fitting to mark the spot. A bronze plaque with their names and "Co. F, 22nd Infantry" and the date is along the lines I have in mind…Do you think any of the fellows in F Co. would be interested in contributing?

In any event, Bill, it was a day of nostalgia second to none in my mind, and I did want to share it with you.

Kathryn and I left Normandy and spent three days in Paris. The City of Light was liberated on August 25, 1944, when the first American troops to enter were from Dad's proud 4th Infantry Division. On the flight home, as she was falling asleep, Kathryn turned to me and asked in a soft voice, "What happened to the men after Pop-Pop left?"

I started to tell her about the battles to seize Cherbourg, but words failed me. I was thinking that a veteran's daughter should respect her father, no matter how she feels about politics. I wished with all my heart that I had apologized to him just once for the words and the assumptions of my youth.

Part Six

A DEEPER UNDERSTANDING

"Dachau and the memory of the experiences of our GIs—the trauma, the heartbreak, the good times, and the really, really bad times—should always be with us."[81]
—**Matthew A. Rozell**, *The Things Our Fathers Saw*, 1961

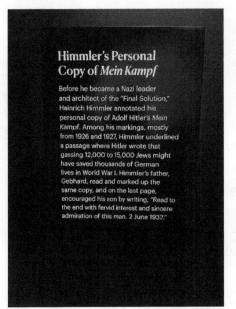 

Museum of Jewish Heritage—A Living Memorial to the Holocaust. 2023 Core exhibition "The Holocaust: What Hate Can Do" Volume I and II of Himmler's Personal Annotated Copy of Mein Kampf, Volume I: In honor of a US infantry officer who helped conscience triumph over Nazism., Volume II: Anonymous donation in special honor of "the girl in the red coat."[82]

26. The Girl in the Red Coat

2019 - 2021

My time with Kathryn from Nice to Normandy had been everything I hoped for: adventures and misadventures, medieval villages and a jazz festival, hilarious moments, and long conversations. The time shared in Normandy had been filled with stories about her grandparents, including special memories of my mother, her beloved Nan, who raised her as much as I did.

In returning to the places where my father had fought, I had discovered a man whose military service I wished I had respected more and had the chance to learn from. I realized now that my father's silence with his family was about combat, not about his war experiences in general. The only way I had been able to learn about the serious engagements his platoon had participated in was through his letters to the men he served with and in Chaplain Boice's history of the regiment. I could better understand why Dad had recounted the stories he told us in the same casual, somewhat ironic tone much of his narrative had.

```
I was the officer in charge of securing
Heinrich Himmler's home at the end of
the war. There was a large, gold embossed
Mein Kampf on a pedestal next to his
```

> desk. It was too large for my pack, so I took the smaller volume from his desk. When I got back to headquarters, I realized it could go home with my commanding officer or me. I decided to keep it.

Most importantly, I had experienced the bravery and death of young fighting men in a personal way that I had not anticipated. To paraphrase Charles Wertenbaker, I had learned there is no "perhaps" about it; we should all know the cost of war. Standing at the Chateau de Fontenay, rage and deep sadness had overwhelmed me as Kathryn and I sought to honor the men who fought and died there. I could not begin to imagine the feelings my father experienced when he returned in 1979. He had called it a "day of nostalgia second to none in my mind."

Somewhere between Utah Beach and the Chateau de Fontenay, F Company, and the Second Battalion of the 22nd Infantry Regiment had taken on a new meaning for me. My father's story and their story had begun to merge. I was affected by how Dad had devoted time in his narrative to acknowledge and describe the men.

Bill Weingart's letter correcting Dad's assumption that Bill unjammed his own carbine during the attack on the Azeville Battery made it clear how teamwork had saved the day. "I exchanged rifles with Ivey on that occasion and he finally cleared the jam. Ivey, if you recall, was a piss poor garrison soldier, but in combat, he got the job done."[83]

I kept returning to Kathryn's question, "What happened to the men after Pop-Pop left?"

The men. Millions of young men called to duty as Dad's fraternity brother had predicted in 1939, "One thing for certain is that you and I will be performers in the biggest show on earth one of these days. I hope I am strong." Where did the "I" end and the "we" begin when confronting something of the magnitude of a world at war? I decided I needed to better understand World War II. Until now, I had focused on D-Day and my trip to Normandy.

Squeezing my curiosity into the time left after the demands of family, work, and pending retirement, I listened to *The Great Courses, World War II: A Military and Social History*, by Professor Thomas Childers of the University of Pennsylvania.

After all my research, I discovered to my dismay that I still possessed only a vague working knowledge of the war. Over the course of many months, as I stole minutes and hours from my schedule, I began to wonder why my schooling in American history had ended just as World War II began. Asking friends, I found another almost universal truth. Not only were most parents silent about the war, but it also seemed the end of American History in high school arrived before there was time to teach their children much about it.

As I listened to the lectures reveal the overwhelming scope of the conflict, my father's voice began to disappear into the dramatic magnitude of his moment in history. Operation Overlord may have been the largest amphibious landing in the history of warfare, but the deadliest battle in the history of humankind had already been fought, in 1942, at Stalingrad in the Soviet Union (now Russia).

To keep things manageable, I decided to consider my father's story within the context of one month—June 1944. The efforts of the 22nd Infantry Regiment's 3,200 combat infantrymen, however critical to the success of the Allied offense, were just one part of a much larger war that was affecting almost every country in the world in some way. Some of the significant battles in June 1944 that I identified were:

June 4: Allied forces entered Rome and German troops fell back to the Trasimene Line.

June 6: The Allies landed in Normandy.

June 9: The Soviets launched an offensive against Finland that led on June 25 to the Battle of Tali-Ihantala, the largest battle ever fought in the Nordic countries.

June 13: Germany launched the first V-1 flying bomb attack on England.

June 15: Marine and Army forces invaded Saipan in the Pacific.

June 18: Free French forces liberated Elba.

June 19–20: The American defeat of the Japanese in the Battle of the Philippine Sea.

June 20: The British continued their advance through Italy and liberated Perugia.

June 21: The Allied offensive in Burma began.

June 22: Operation Bagration, the Soviet offensive in what's now Belarus, began.

June 25–26: The Allies captured Cherbourg in Normandy.

And where did all these violent attacks and counterattacks by great armies get the world?

I will let Dr. Childers summarize:

> 55 million people would perish across the globe. [Now estimated at 75 million or more.] No corner of the world would be left untouched. Death on a scale never before imagined. These numbers are staggering, impossible to comprehend. And yet those human costs, I think, are things we must struggle to understand, must come to terms with. The numbers in many ways leave us blank. They are too big, too staggering. Individual stories, I think, bring the meaning of this war much closer to home.[84]

Individual stories. I thought of the Chateau de Fontenay where my father stood twice, and Kathryn and I stood once. Wasn't it likely that for my father, the young men he knew and lost there would forever represent those 55 million lives? I hadn't so fully understood before how the power of history can lie in the stories of its participants. I began to look for published accounts by veterans from the 22nd Infantry Regiment who fought in World War II.

And then, in the spring of 2020, I experienced my own global upheaval when the world was plunged into the deadly COVID-19 pandemic. As life assumed new dimensions of fear, death, and restrictions, I continued my research but completely overlooked an exhibit of great personal importance. A letter addressed to Mom that had been misplaced earlier went unread until the spring of 2021. It informed her that the copy of *Mein Kampf* she had donated to the Museum of Jewish Heritage had been chosen for display in a new exhibition, "Auschwitz, Not Long Ago, Not Far Away." This was a comprehensive exhibit curated by highly respected Holocaust scholars.

The letter thanked my mother and indicated that when the

exhibition opened on May 8, 2019, it heralded a new era for the museum. This was the first correspondence we had received since the donation in 2006. Finally, I contacted the museum on May 1, 2021, two days before the exhibition closed, to let them know my mother had died and to see if Hunt and I could attend. It was sold out, but they made an exception.

The next day, my husband and I made our first trip to New York City since the pandemic had begun. The anticipation of seeing the book on display mingled with the peculiar sensation of enjoying formerly mundane events, like finding a parking garage. As we approached the museum, our lighthearted mood changed. The sky seemed foreboding, somber gray clouds hovered over the Hudson River and the museum.

Before we left for New York, I had read several essays that recounted how Heinrich Himmler became one of the most powerful and influential men serving Adolf Hitler. From 1923 to 1929, his failure to establish a successful career paralleled his growing involvement in paramilitary, nationalist activities and his affiliation with the Nazi Party. In 1929, Hitler appointed Himmler Reichsfuhrer of the Schutzstaffel (SS). By the time the Nazis came to power in 1933, Himmler had grown the SS to 52,000 (from 280). Himmler inducted two new functions for the SS (who were initially Hitler's personal bodyguards)—internal security and guardianship over racial purity.

> Following the Nazis' invasion of Poland in 1939, Hitler appointed Himmler as the Reich Commissar for the Strengthening of the German Ethnic Stock. This new position authorized Himmler and the SS to have complete control over German resettlement in areas of occupied Poland and eventually the Soviet Union in 1941.
>
> ...Within the history of World War II, Himmler is best

FINDING MY FATHER'S FOOTSTEPS | 235

known for his leadership role in the implementation of the Nazi concentration camp system and the Holocaust. In 1933, Himmler established the Nazis' first concentration camp in an abandoned munitions factory at Dachau...By the end of the war, the Nazi camp system grew to contain thousands of camps, including concentration camps, forced labor camps, POW camps, transit camps, and killing centers.

...Himmler was captured by the Russians on May 20, 1945, and committed suicide in British custody on May 23, 1945.[85]

A cattle car used to transport Jews and political prisoners to their deaths stood outside the museum. Hunt and I paused to read the interpretive panel. It depicted a crowd of people with a young woman standing in front, she was crying as she clutched her hands in fear. The inscription said, "A young Jewish woman from the Greek town of Joannina during the roundup of March 25, 1944."[86]

Entering the exhibition, we saw delicate lights embedded in a black background with golden words that read: "Dedicated to those of other faiths who risked — and often gave — their lives to save Jews during the Holocaust."[87]

My thoughts went to the men of the 22nd Infantry and my father. I felt a rush of pride. The *Mein Kampf* they secured was contributing to this important exhibit.

As we walked into the first room, there was a large picture on the wall. It will forever be emblazoned in my memory. It showed a pile of thousands of shoes with a single, now ruined, dress shoe sitting toward the top. I thought of Dr. Childers' words, "Death on an incomprehensible scale. Human cruelty at its worst."[88]

The next portion of the exhibit traced the rise of the Third Reich. My father's volume of Heinrich Himmler's *Mein Kampf* was displayed as a historic artifact that reflected Hitler's philosophy.

I read our family's proud dedication:

> *In honor of a US infantry officer who helped conscience triumph over Nazism.*

It was left for the visitor to ponder how words were translated into reality in the hands of that book's owner. We moved silently through the next rooms, learning about the evolution of the Final Solution, the building of the Auschwitz concentration and extermination camp complex, the propaganda of hatred, and the implementation of mass murder which became fanatically efficient and deadly after Germany's invasion of the Soviet Union in 1941.

A second lieutenant at the Belzec killing center described how victims were whipped into submission as they got off the trains, then were stripped of clothing, hair, eyeglasses, and artificial limbs. As the naked men, women and children moved toward the gas chambers, "…a strong SS man stands who, with a voice like a pastor, says to the poor people, 'There is not the least chance that something will happen to you!' The chambers fill. 'Pack well!' Captain Wirth orders. The people stand on each other's feet. The doors close."[89]

I stopped before a picture of a lovely young girl with a group of women—destined to be transported to their deaths. Was this the girl the Williams' family dedicated Volume II to?

> *Anonymous donation in special honor of "the girl in the red coat."*

My heart ached.

The last room featured interviews with concentration and death camp survivors. They spoke of the strength to live found in forgiveness, but also of the imperative to always remember. For me, the exhibit had ended with empty shoes and a little girl.

Hunt and I were speechless as we left the museum. I gave thanks with all my heart to the person who had selected the beautiful site where the museum stands. The eternal rhythms of water and sky were about the only sensory experiences I could endure. I wondered if the architect understood museum visitors would feel as I did. I was trembling. I could not forget the shoes and their owners. I had taken a picture of a poem written by an Auschwitz survivor that was highlighted in the exhibit. I read it again.

This dot on the map
This black spot at the core of Europe
this red spot
this spot of fire this spot of soot
this spot of blood this spot of ashes
for millions
a nameless place.
From all the countries of Europe
from all the points on the horizon
trains converged
toward the nameless place
loaded with millions of humans
poured out there unknowing of where
poured out with their lives
memories
small aches

huge astonishment
eyes questioning
bamboozled
under fire
burned
without knowing
where they were.
Today people know
have known for several years
that this dot on the map
is Auschwitz.
This much they know
as for the rest
they think they know.

I realized that when Charles Wertenberger wrote about the politicians', generals', and soldiers' wars, he had left something out. He had not included the civilians' war. And he had not mentioned atrocities. He only alluded to the delicate line between humanity and savagery. I wondered if there was a serpent in the heart of every human waiting to be unleashed by war and hatred. Or if only some of us were so cursed.

As I looked out at the Hudson, I decided that war was an experience like no other thing that a person, soldier, or civilian, ever faces. The rest of us can only bear witness to its horrors. We should resist the temptation to think "we know."

55–75 million lives lost worldwide. How many more lives scarred or ruined?

Could my parents' silence about the war years have been their way of trying to protect me from what they had experienced?

Am I getting any closer, Dad?

27. Pieces of the Puzzle

2021

It was September, four months after Hunt and I had been overwhelmed by the exhibit on the Holocaust at the Museum of Jewish Heritage. I was standing in the sunny, familiar nave of Norton Presbyterian Church in Darien, Connecticut. Looking out through eyes dimmed by tears, I saw the faces of my family and childhood friends. I was trying to celebrate the life of my sister Elizabeth while consumed with rage over the cancer that had returned and destroyed her body in those four hellish months.

Home again, I was sitting on the yellow love seat, looking up towards my father's shelf in desperate need of comfort from my parents. I had retired almost simultaneously with the return, after fifteen years, of Elizabeth's cancer in late April 2021. My world, already crumbling from her devastating disease, had been further shaken by my recurring worries about the *Mein Kampf* on exhibition at the museum. Was our family's proud dedication tainted by how the book was acquired?

As I had done once before, in the face of divorce, I undertook a research project to try to steady myself through an emotional storm. Seventeen years after my first conversation with Mr. Williams and almost thirty years after my father's death, I decided to see if I could finally uncover the details of my father's war record after his narrative ended in January 1945. I wanted to know how, when, and why he was at Heinrich Himmler's home at the end of the war. I needed his version of the *Mein Kampf* story to be true. Fate had stolen my sister; I would force it to return my father as I knew and loved him. I would prove him innocent.

After much thought, Dr. Breitman's words reminded me that my first obligation was to the truth.

> The real historian can and should describe and analyze events to make it difficult for others, like Himmler, to wear blinders when they gaze upon the past and fashion their own myths. Historians cannot demythologize the past if they are willingly blind to inconvenient facts that do not fit into their grand theories.[90]

I understood that people tend to alter facts to suit their own needs. I knew that by May 1945, my father must have been weary of things no one should ever endure. Was his story just one more manipulation of a difficult script? Reluctantly, I set my feelings aside and decided to see where the facts led me, hoping they would corroborate his story.

My first step was to determine where Heinrich Himmler lived in Germany and, if he had several residences, identify the one that played a role in my father's story. Once that was established, I hoped I could find records that would help me trace Dad's whereabouts after he returned to Europe in 1945. I would follow in his footsteps to see if they led to Himmler's home.

I found a lone website with a photograph of Heinrich Himmler's residence, captioned:

> Many of the National Socialist elite had homes around the Tegern-

see. On the left is a May 1945 view of SS Reichsführer Heinrich Himmler's house "Lindenfycht" near the lake shore in Gmund.[91]

There were many mentions of another Himmler property, Wewelsburg Castle, which had its own strange story, unrelated to mine.

Gmund am Tegernsee was easy to locate; it was near the Bavarian Alps. Then I discovered a photograph online of soldiers in Himmler's home, listening to the radio on V-E Day. They were a Texas regiment, the 143rd Infantry Regiment of the 36th Infantry Division. There was no sign or mention of the 22nd Infantry Regiment.

I looked around my cluttered office. Was it all a lie? Did my father end the war in an office in Paris? No, I told myself, I was being ridiculous. There was nothing in his files to indicate he did not finish the war with the 22nd Infantry and Chaplain Boice had placed the regiment in Bavaria. The only question was if his story about getting the book was just that, a story.

I picked up the 4th Infantry Division Yearbook roster published in September 1946 and walked out on our back deck. I went through the yearbook page by page until, finally, just below the picture on page 138 of Major Clifford Swede Henley, Commanding Officer, Headquarters, Second Battalion was one with the smiling face of First Lieutenant John Sisson, S-2, Intelligence Officer.

Mr. Williams' question assailed me. "Was he in intelligence?" Why hadn't I noticed this before? Dad had indicated in his narrative that he had held that position as of late October 1944. Could the book have come into my father's possession while he was the S-2, and he took it? Just as Williams had accused him of doing? But why tell his elaborate story? My anxious confidence collapsed.

It took many days, but curiosity and dogged faith in Dad got me back at my desk. Getting the Second Battalion Headquarters and my father to Bavaria and Tegernsee ahead of the 36th Infantry Division was my next job. By its own admission, the 4th Infantry Division did not play a lead role in Europe as the war was ending. Even the souvenir roster I had consulted only mapped their advance from Utah Beach to the fighting at Germany's Siegfried Line, in the Hürtgen Forest and in the defense of Luxembourg during the Battle of the Bulge. I would have trouble locating the answer in general war histories.

It was time to organize my father's papers to see if I could find the evidence I needed. I had already done a fairly good job with his files back in 2016, but the process still took several months. Fortunately, the old lawyer kept the pertinent documents:

- Chaplain Boice's history of the 22nd Infantry Regiment, 4th Infantry Division.
- My father's edited story and war records.
- My father's photo albums from 1944 to 1945.
- My father's 22nd Infantry Regimental Society Newsletters in chronological order.
- My father's correspondence with his Army peers.
- My father's letters home to his parents from 1942 to 1945 in chronological order.

Most importantly, I identified Dad's military "personnel jacket." It was the typed form that provided the official dates and assignments for 2nd Lt. Sisson from 1943 to Major Sisson in 1951. This was what I had needed all along. I found he had traveled back to Europe from February 27 to March 10 and awaited reassignment from March 11 until April 5. He reported for duty to Major Henley on April 6 and got his old job back at 2nd Battalion Headquarters as S-2 with a brief assignment of Adjutant from May 6 to May 12. The war ended for my father on September 4, 1945, when he headed home for good.

Why didn't I do this years ago?

I could now read Chaplain Boice's history with confidence that my father was with his unit after April 6. The history placed the 22nd Infantry Regiment's Second Battalion at Holzkirchen, Germany on May 2. This was about 20 kilometers from Gmund am Tegernsee. Boice did not say where the battalion's headquarters were located.

I must pause at this point in my story to say to anyone who might be inspired to consider researching their own family stories, don't hesitate to begin. And don't worry about getting it all correct the first time around. Somehow, I misread Chaplain Boice at this point in my research. This did not become apparent to me until much later in my project.

From the newsletters, I learned a great deal of seemingly unrelated but helpful information about my father's relationship with his regiment after the war. The 22nd Infantry Regiment Society was formed in 1945 to maintain the fellowships formed during the war years. The importance of keeping the soldiers in touch with each other was brought home by a moving letter my father received in 1975:

It was a real pleasure to receive your letter just a few minutes ago. My brother, Floyd Robert West, age 18, went into Utah Beach on the early morning of June 7, 1944 as a replacement troop. He was attached to the 22nd Infantry Regiment, though I'm not absolutely sure of the battalion or the company…He was killed June 20, at the beginning of the final assault on Cherbourg. Chaplain Bill Boice wrote me two or three times about him at the time of his death."[92]

The writer, Earl West, had visited the Chateau de Fontenay and shared pictures from his visit with Chaplain Boice, who published them in the newsletter. My father had inquired about the pictures.

Earl West closed his letter by saying, "I figure that my brother wrote his last letter to us from the chateau. That is why I was interested in it and took pictures of it. I can still see it in my mind's eye." I don't think it was an accident that four years later, my father made his own journey of remembrance there.

In 1980, Dad wrote to Chaplain Boice about his visit and made a special request:

> Dear Bill,
>
> I did have a chance last October—AWOL again, this time from business—to drive from Cherbourg to Utah Beach and then try all the back roads to trace F. Co.'s route from the beach to the Chateau Fontenay. Couldn't put it all together, even with a tactical map I've been hanging onto for years, but Azeville and the Chateau came back with particular clarity.
>
> When I finally oriented the attack on the Chateau and walked the ground, the surge of nostalgia and melancholy involving F Co's first bad bloodying was indeed great. It would be a good thing to place an F. Co. plaque on the road at the entrance to the Chateau in memory of Fulton, Cook, Wirtzberger, and the others

who didn't come out...I would like to hear from any F. Co. old timers who might want to participate if and when anything comes of it. Bless you for your continuing act of faith on behalf of the 22nd.[93]

Chaplain Boice placed an article about my father's trip and his idea for a plaque in the newsletter and a number of men sent letters supporting the idea. There was, however, no indication that a memorial was erected. What did happen was an Austrian soldier who fought on the other side ended up contacting my father, but I could not locate their correspondence.

I enjoyed the exchanges and camaraderie, but I was no closer to my goal until I discovered a letter written October 26, 1990, four months before Dad's death.

Dear John,

I was sorry to read of your health problems in the 22 Dispatch. Sure hope you are on the road to recovery. I had hoped to see you at one of the Fla Chapter meetings, but guess you couldn't make it.

Perhaps you don't remember me, because I was one of many replacements, but I remember you very well. It seems you were always just returning from the hospital or a trip home due to your multiple injuries. I'm sure you were highly regarded at 2nd Bn. Hq. because they always had a job for you. Sorry I didn't get to know you better.

Here's wishing you a speedy recovery.
Regards, George Wilson[94]

Ever the lawyer, my father kept a carbon copy of the reply he wrote on November 2,

```
Dear George,

Thanks so much for your recent letter. It
was a most pleasant surprise and much ap-
preciated.
    You are right. We weren't around 2nd
Bn. at the same time. At the war's end, I
remember you getting H Company and Pil-
lard coming to Bn. Hq. where I worked
with him fairly closely—mostly after the
action was over.
    I'll tell you one thing, George—you did
one helluva job on your book. I've read
a lot of WWII history, and have no hesi-
tancy in putting "If You Survive" in the
front rank of combat description and the
plain humanity (or inhumanity) of war.
It is also, of course, almost a person-
al memento to me. The many names I knew
so well and felt so close to. The place
names and events along the way…You even
had Carl Servat in the act. Servat had a
BAR in the 3rd platoon of F Co. on D-Day
and may have got the first Kraut blood for
the platoon when he blasted a German mo-
tor convoy. And there were so many others
you mentioned.
    I must say that you have every reason
```

to remember me. I felt some of the short hairs crawling when I read your description of going from S-2 to F Co. I had just returned from hospitalization, and I was the guy who was traded for your S-2 spot—a fact I never knew, but of which you were undoubtedly aware. Newcomb and I used to buddy around in England a little bit as unwashed 2nd Johns and he knew I'd had about 4 or 5 days in combat in Normandy while you'd done most of the fight across France. Some coincidence.

    I had hoped to meet you in Jackson, but it didn't work. I'm getting over surgery on my thigh to remove a tumor which apparently was successful, but which has me laid up (still getting hospitalized, but no Purple Heart.) So, I don't know whether I'll get to the next reunion, but I'll try.

Thanks again for your thoughtful letter.

Sincerely,
John[95]

I could not believe my good fortune. I ordered the book from Amazon and read it. Many of the names and places in my father's story came to vivid life, but George Wilson was wounded in February 1945 and did not return to the 22nd Infantry until after the war ended. His memoir could not help me get to Gmund am

Tegernsee, but I now had my third first person narrative from the Second Battalion: Sisson, Boice, and Wilson.

I called my brother. He told me he had no additional files, so I made one last fraught review of Dad's infantry correspondence. And there, printed on blue carbon paper as if to signal its importance, was my smoking gun. A note on a smaller sheet of notepad paper was attached. How had I missed it?

Date: 1 June 1982
To: John Sisson
From: John Ruggles

Please advise me if your address should be changed from Delaware to Florida. Hope to have a good turnout at Callaway Gardens. We will miss you.[96]

My father had typed his response:

July 1, 1982

Maj. Gen. John F. Ruggles
c/o 22nd Infantry Society

Dear Gen. Ruggles:

Please be advised that my correct address is as stated above. I moved from Wilmington, DE, in 1980 and am now safely ensconced in balmy Florida. I realize that you Arizonans don't really believe people in their right minds would move down to

> this swampy area when the hills of Phoenix beckon, but we do avoid scorpions that way.
>
> There is one thing I was going to check with you at Calloway Gardens this summer. I have always wanted to obtain some sort of authentication of the copy of Mein Kampf which I liberated at Heinrich Himmler's summer home on the shores of the Tegernsee (or whichever "See" it was) a day or two before the 22nd was taken out of action in May, 1945. We spoke briefly about it at Asheville, and it occurred to me that your authentication, to the extent of your recollection, might be useful to have. You may recall that the volume was signed by "Heinrich Himmler—1925". I had intended to bring it to the reunion, until it developed that I would not be able to attend.
>
> Regards to Mrs. Ruggles and others you see there.[97]

I was speechless. The family fable had been written down. Just as Dad always said. He was in Himmler's residence on the Tegernsee and was busy "liberating" things near the end of the war. I couldn't help noting my father's formal response to retired General Ruggles' casual note, thirty-seven years after his own service in the army had ended. Clearly, his respect had no time limit.

I spent several days celebrating—and then I stopped. Even as an amateur historian, I was all too aware that being close proved nothing. Although now it seemed unlikely, Mr. William's assertion still bothered me. Was my father's letter repeating a story that was only partially true?

The historian's job is to sift through the debris of memory, documents, and artifacts to assemble an account of past events that is as free of opinion and bias as possible. I needed more information to place my father where he said he was, in Gmund am Tegernsee in early May 1945.

28. We Shall Remember

WINTER 2022

"They went with songs to the battle, they were young,
Straight of limb, true of eye, steady and aglow.
They were staunch to the end against odds uncounted.
They fell with their faces to the foe."[98]
  —**Laurence Binyon**, For the Fallen, 1914

In my quest to finish Dad's story, I was now reading the first-person memoirs of infantrymen and books that featured interviews with servicemen who had fought in Europe during World War II. I had not found anyone who could place Dad in Gmund, but I now saw his story as less personal and more universal. Sometimes the pain of my sister's death and that of the events the men described was overwhelming.

I happened to visit an old friend in Vermont the winter following the exhibit at the Museum of Jewish Heritage. Aware of my project, she made a point of showing me a framed hand-drawn map that her husband had proudly hung in his study.

My friend's father-in-law, Albert Goetze, Sr., was a sergeant and squad leader with the 334th Infantry, 84th Division in Europe during the war. He had indicated on the map the towns where he fought during the Battle of the Bulge. As I looked at the map, I noticed a stanza from a poem that he had attached to it. The stanza was from Laurence Binyon's famous World War I poem, "For the Fallen." Mr. Goetze had altered the stanza:

> *They shall not grow old*
> *As we that are left grow old*
> *Age shall not weary them*
> *Neither the years condemn*
> *But we shall remember them.*

> *WE SHALL REMEMBER THEM*[99]

And then I noticed the faded black-and-white photograph of five smiling soldiers with the following typed caption:

Standing (l. to r.)
    Pfc. Patrick S. Dillon, Jr.   Killed in Action
    Pfc. Lawrence T. Benson   Killed in Action
    T/Sgt. Albert F. Goetze   Wounded in Action

Kneeling (l. to r.)
    Pfc. John Adair   Wounded in Action
    Pfc. David Carson   Killed in Action

The somber words and engaging faces of the young men made me realize that the map possessed a far deeper meaning than I had first supposed. I was standing before a survivor's memorial to his fallen comrades.

The filing cabinet in my parents' attic. The unshared journal and the medals quietly stored. My father's letter about his return to Normandy and his desire to memorialize his fallen comrades at the chateau. These told his story of courage summoned, and of life lost and barely kept. Like the map before me, did they too stand in silent remembrance?

Familiar tears came to my eyes as I again pondered the politicians' war, the generals' war, and the soldiers' war. I had spent most of my life confusing them. On my cell phone I read the poem memorializing the young lives lost in World War I in its entirety. I was glad my friend had left me alone. Within its measured stanzas, I found words that helped give voice to the complex feelings I was trying to understand as I pursued my father's involvement in World War II.

I repeated the last line from the third stanza aloud, "They fell with their faces to the foe." One of the articles I had read was written by 22nd Infantry Regiment staff sergeant David Roderick. "In World War II in Europe when your outfit went on the line you stayed there. It was just a matter of time before you were seriously wounded or killed."[100]

Standing before the map, I thought about how every soldier's story I read spoke of two things—finding the courage to remain staunch against the unimaginable odds and honoring those who fell facing the foe. I kept finding references of passing the frozen or bullet riddled body of a nameless soldier—with the unspoken understanding that the author could be next.

Men tried to describe the physical and emotional exhaustion and terror of holding their ground or advancing while mortars and bullets exploded with mind-numbing noise around them. The anguish as they witnessed colleagues being killed or injured, some having just joined them as replacements. I thought about the young men who returned home to face the foes that tormented their sleep and memories. Many men survived the war and prospered, but others paid a terrible living price.

My neighbor Ed, who was seriously injured in Vietnam, said to me, "It sounds weird, but I was almost relieved when I realized I had been seriously injured. I was afraid if I stayed much longer, I would lose my humanity." Was it the attempt to maintain humanity in the face of atrocity and death that was behind Dad's accounts of his escapades? Was he struggling to put a human face on the inhumanity of war? But the overwhelming and unspeakable reality of war lurked in his narrative.

Dad returned to the chateau to remember and honor those who fell during F Company's "first bad bloodying." It was one and the same as Mr. Goetze's memorial before me.

But where our desires are and our hopes profound,
Felt as a well-spring that is hidden from sight,
To the innermost heart of their own land they are known
As the stars are known to the Night... [101]

After I got home from Vermont, I returned to one of my favorite memoirs, Sgt. James Cullen's *Band of Strangers*. He vividly described the terrible conditions he and Mr. Goetze had fought under. The chapter entitled "The Valiant Volunteer" recounted an engagement in the frigid Ardennes during the Battle of the Bulge.

> Firing started, it was all incoming...Mortars and heavier shells slammed into the snow...One guy dropped in front of me. He was curled up, so I could see his face. He was shaking, and his face was quivering with his eyes tightly shut. He was in a state of near panic. In that hell of a situation, we were all deathly afraid, but we hung in there. We had to. Our job was to take the town up ahead.[102]

FINDING MY FATHER'S FOOTSTEPS | 255

After they took the town, the men were cut off from the American lines. Cullen saw a group of soldiers cowering in a stone basement while he and others were fighting off the vicious German attacks. An unnamed young tanker volunteered to head into the frozen fog to try to make contact with the American lines. Cullen notes that just after the kid took off, "I looked back into the room to check on the goldbrickers, and they were still squatting where they were before."[103]

Cullen found a spot where he could watch for the young tanker:

> I saw him coming out of the fog. He was staggering through the deep snow and stopped near the back of the C.P. to climb over a wire fence.
>
> As he stopped, a shot rang out from below my window, and the kid fell and didn't move. Those lazy bastards in the room downstairs had shot him.
>
> I felt like firing at them through the floor. I nearly cried. That brave young tanker had been shot by some yellow bastard. What an idiot, and what a waste.[104]

The first time I read the passage I felt physically sick.

What was courage and honor in the face of mayhem and terror? What constitutes cowardice versus panic? History books rarely discuss this. The soldier lived it, saw it, fought it within himself and tried to encourage others. Knowing you were terrified, not brave, but managing to face the foe — or not. Justifiably outraged, Cullen's first impulse was to fire through the floor, but he did not. How did he live with what he had witnessed?

I remembered my father's description of his D-Day landing:

```
Suddenly there were no more questions.
```

> Only a G.I. saying, "those b------- are trying to kill us sure as hell." That is all that is needed to make a soldier fight. That is what makes him do brave things or cowardly things. Just that one phrase. There is no motive or thought behind a trigger squeeze.

At 95, after many years pondering his experiences as a young soldier, James Cullen offered me a profound warning. It is always dangerous to glorify, in any way, the soldiers' war. He addressed the question of courage and cowardice and the price paid by fellow soldiers in a way that changed how I thought about combat. Honor played a pivotal role on the battlefield, but how could an individual soldier be held accountable unless it was by his peers. At the Museum of Jewish Heritage, I had realized that the civilians' war was also part of the soldiers' war. Sometimes victim, sometimes enemy, and sometimes fellow warrior. And the profound caution to never assume that "I know" what that means.

The soldiers' war. That was the history I wanted to try to understand. It was created by politicians and managed by generals but fought by soldiers—still boys or barely men. Boys that Hitler was banking on not being able to make the grade. And yet, just like my generation's Vietnam vets, they each responded as best they could. And their best was way better than expected in most cases.

It seemed to me that in the soldiers' war there was a wavering, not always obvious line between action and terror, honor and atrocity, courage and cowardice. Only those who served knew the truth about who managed to keep their face to the foe and who did not. And those who served saw the consequences when humanity was lost. Only those who served knew the foe was both without and within. They understood the price of honor. I wondered if this was part of the reason no soldier returns home with much to say about their experience to those who weren't there.

Those who weren't there. The file drawer with all the unshared stories. I felt a little foolish when I finally realized that my father's story was one with his comrades, not his family. If my dad fought his way to Himmler's residence, it wasn't by himself. His story could not be finished without his battalion and his regiment. It was just like F Company; he and they were somehow one and the same. Their success was his success and their failures his to share.

I made a crucial decision, which would add a year of research to my journey. I would follow the men of the 22nd Infantry Regiment from Utah Beach to Bavaria and see where we ended up. I wanted to see if it was possible to begin to feel what they felt; to penetrate the frustrating and confusing silence of their generation when it came to World War II. I did not want to be protected from it anymore. And, maybe, in doing so, I could find something that would put my last questions about my father's actions at the end of the war to rest once and for all.

One soldier and a few hours in Heinrich Himmler's library in the spring of 1945. I was looking for a needle in a haystack. I had a few eyewitnesses. I needed more.

29. Deeds Not Words

2022

Fifty-three years after I first demonstrated against the Vietnam War, I finally turned to those I had seen as the enemy for answers. I contacted Bob Babcock, the historian and president of my father's old outfit, the 4th Infantry Division Association. I was done with climbing the mountain backwards. Babcock was a veteran, and I was a veteran's daughter who needed his help.

I had come to realize that my adage that all war was bad and should be avoided was simplistic at best and dangerous at worst. Elliot Richardson's address to the Twenty-Second Infantry Regiment Society Reunion in 1973 offered an intriguing observation. Richardson had been a medic with the 22nd on D-Day 1944. During the Vietnam War, he had served as both Secretary of Defense and Attorney General in 1973 under President Nixon. He would resign in October of that year when Nixon fired the Watergate special prosecutor Richardson had appointed.

> We didn't have an elite unit, but there was something special about the spirit of the 4th. Some units had it; some didn't...We knew that the things we valued most could well be destroyed unless we were

willing to make the sacrifice…That has been less clear perhaps in Vietnam, but it is true, nevertheless.[105]

I had no regrets about my active protests against a war I believed our country had escalated without cause. But I had failed to respect my father and the soldiers' war in the process. As Richardson pointed out, the men of the 22nd Infantry Regiment were making the sacrifice to protect my freedom in World War II and again in Vietnam. They were not making the decisions.

On May 4, 2022, in answer to my query, I received the following email from Bob:

> Great to hear from you, Karen. You hit the jackpot by contacting me. I probably know as much or more about the 4ID [4th Infantry Division] and 22nd Infantry Regiment in WWII (and after) than most people living today. Not bragging, just am proud of having been the 4ID historian since 1998.
>
> I would be happy to talk to you at your convenience. There was never a post WWII history of the 4ID written. There are lots of books out there. Below I am attaching a Word document with all the 4ID books I'm aware of.
>
> The only complete 22nd history from WWII is the one Dr. Bill Boice wrote in the mid-1950s. I also have that in digital format.
>
> Another treasure I have is a daily After Action Report (AAR) that a WWII vet got from the National Archives and gave me. Those were transcribed by Philippe Cornil and I have that in digital format as well. As you read the daily narrative, it is always in the format of overall 4ID statement first, followed in order by 8th, 12th, 22nd Infantry Regiments—so 22nd is always at the end of each daily report.[106]

A born historian, Babcock wrote a term paper about D-Day in high school, never knowing he would serve as an officer with B Company, First Battalion, 22nd Infantry Regiment of 4ID from 1965 to 1967, including Vietnam. He got to know the men from my father's past when he attended his first National 4th Infantry (IVY) Division Association meeting in the early 1990s. Many members of the 22nd Infantry Regiment Society were also members of the National 4th Infantry Division Association. Bob has served ten years as president of both organizations over the past 25 years. He had become friends with George Wilson, John Ruggles, Bill Boice, and others before they died.

Bob Babcock was an inspiration and became my guide to learning about the 22nd Infantry Regiment. That is, after our first telephone call when I broached the subject of our very different Vietnam Wars. He served—I had called our servicemen names. In his generous and forthright manner, he said, "I don't let those things trouble me. Everyone must live with his or her conscience."

Thanks to Babcock's careful tending of division history, I was able to accomplish my goal of mapping the weekly, sometimes daily, 1,600-mile advance of the 22nd Infantry Regiment through Europe to see if we would end our journey in Gmund am Tegernsee. My focus was on the Second Battalion, but as I turned that journey into Part Seven of my book, I also included additional stories for context.

The men of the 22nd were not elite troops; they were the hard-working infantry. They landed at the beach that wasn't infamous, and they ended the war in Bavaria, not Berlin. Personalities, tanks, and aircraft upstaged them at every turn, even though they often spearheaded the attack. But they were men who fought against man and machine on the front lines of combat almost without respite from June 6, 1944, to May 8, 1945. David Roderick summed it up:

> In terms of sheer hours of infantry fighting, the European Theater contained 68 combat divisions. The most casualties suffered were by the 3rd Infantry Division that fought in Africa, Sicily, and Italy. The 4th Infantry Division was second.[107]

What they might have lacked in publicized glory, they more than made up for in endurance and courage. I believe that within the story of the IVY Division (the proud nickname of the 4th—IV—Infantry Division) is the story of all the men and women of their generation who were called upon to serve their country, whether at home or abroad. I am not alone in my respect and affection. Ernest Hemingway no less decided that the men of the 22nd Infantry Regiment were the guys he wanted to report on and, to the distress of some in command, hang out with.

What follows are moments in the American advance from France's Normandy beaches toward Germany's Bavarian Alps as told by a few of the soldiers who walked, fought, and rode all or part of the way.

These are the first-person voices from the 22nd Infantry Regiment; all but George Wilson landed on D-Day:

- John Sisson, 2nd Lieutenant, Third Platoon, F Company; 1st Lieutenant, S-2 Second Battalion Headquarters.
- Dr. William S. Boice, Captain and Chaplain, 22nd Infantry Regiment.
- Clifford "Swede" Henley, Captain/Major, C.O. Third, First, and Second Battalions.
- George Wilson, 2nd Lieutenant, E and F Companies;

> 1st Lieutenant, S-2 Second Battalion Headquarters, C.O. Company H, Second Battalion.

John Sisson, my dad, has already been introduced.

Bill Boice was born in Blackfoot, Idaho on December 21, 1915. He graduated from Cincinnati Bible Seminary, Cincinnati Conservatory of Music, and Harvard Chaplain's School. He served as chaplain to the 4th Infantry Division's 22nd Infantry Regiment from August 1943 to February 1946, landing with the first wave at Utah Beach on D-Day. He convinced the Regimental Commander, Colonel John Ruggles, on the ship coming home from Europe in July 1945 to form the 22nd Infantry Regiment Society. He completed the *History of the Twenty-Second United States Infantry* in 1959. It has recently been published by Deeds Publishing.

Swede Henley hailed from South Carolina, landed on D-Day, and was never wounded. His roommate in basic training said, "I never knew of anyone who wasn't his friend or didn't like him. The only possible exception to this would have been the Krauts, but they didn't know him personally—at least the way we did. You can bet your bottom dollar though, those Krauts that did get to know him, respected the hell out of him. Swede left his mark at Clemson the same way he left his mark with the 22nd Infantry. He was one helluva football player, as well as an excellent boxer. Naturally, he was one of the best-liked men on campus." Henley maintained a diary while he fought in Europe (not authorized), which I carefully followed.

George Wilson, like Dad, was a Midwesterner. Hailing from Michigan, he was a high school football star. He landed at Omaha Beach in early July 1944, and served in combat on the front line continuously for eight months as a lieutenant with the Second Battalion before he was wounded. He wrote about his experiences in *If You Survive,* which was required reading at West Point for many years.

Their voices are joined by those of a number of men who served or were war correspondents with them. Unless indicated, each man landed on D-Day and served with the 22nd Infantry Regiment:
- Charles T. Lanham, Colonel, Regimental Commanding Officer, July 9, 1944—March 2, 1945.
- John F. Ruggles, Colonel, Executive Officer 1943 to March 2, 1945, and Regimental Commanding Officer, March 3, 1945—February 1946.
- George Bridgeman, Lieutenant, Third Battalion.
- Earl "Lum" Edwards, Colonel, C.O. Second Battalion and Regimental S-3.
- Donald Faulkner, Captain, C.O. E Company, Second Battalion, September 1944.
- John Groth, *Parade Magazine* illustrator, war correspondent, specific reports on the 22nd Infantry Regiment.
- David Roderick, Squad Sergeant, H Company, Second Battalion, 81mm mortar.
- Stan Tarkenton, PFC Machine Gunner, M Company, Third Battalion.
- Charles Wertenbaker, *Life* magazine reporter with the 4th Infantry Division.

Staff Sergeant Jim Laferla's story is told by his nephew and Vietnam veteran, James Davis, as recounted to him by Micky Lieberman, Medic, and Gerald Crane, Staff Sergeant. Both men served with Jim in the Second Platoon, H Company. There are other soldiers who served with the 22nd Infantry Regiment. I have introduced them when they are quoted.

I have also included a number of accounts from soldiers who crossed paths with the regiment. Chief among them are Squad Sergeant James Cullen and Private John Cassar.

Cullen landed on Omaha Beach in the first week of July. He served with the 3rd Armored Division, 36th Armored Infantry Regiment as a replacement Staff Sergeant. He was briefly attached to the 4th Infantry Division during the Saint-Lô Breakout. His memoir, *A Band of Strangers*, gave me a bridge to the soldiers' war.

Cassar arrived on Omaha Beach with the 7th Armored Division in mid-August 1944. The "kid from NYC who never owned a car" became a truck driver delivering ammunition and fuel to tanks. His unpublished diary was made available to me by his son. The 7th was initially assigned to the Third Army, commanded by Gen. George S. Patton, Jr., but served with the First Army after mid-October according to Cassar's diary.

It is difficult to tell the soldier's story without their generals. Gen. Omar N. Bradley commanded the 12th Army Group, which was comprised of forty-three Divisions and approximately 1.3 million men. The 4th Infantry Division landed at Utah Beach under Bradley's First Army, commanded by Gen. Courtney H. Hodges, and finished the war with the Third Army, under Gen. George S. Patton, Jr.

One voice that I am not including, but which can be heard in the background is that of Lt. Col. Arthur S. Teague. He landed at Utah Beach and commanded the Third Battalion of the 22nd Infantry Regiment throughout the war. He became the regiment's Executive Officer following John Ruggles. I now understand why Dad had been so excited to have his family ride on the Mount Washington Cog Railroad. This wartime commander ran it![108]

What the soldiers all had in common was they were the *Poor Bloody Infantry*. Each man lost a highly esteemed friend, and many were wounded in battle. Jim Laferla was killed. Each experienced the brutal reality and terror of front-line combat. They slept in rain-filled foxholes and fought in unimaginably bitter cold. They wrote many letters home, swore liberally, and strove to maintain their humanity.

Tracing my father's footsteps in Normandy got the men to June 10, 1944, and through my earlier research I had continued the story to Cherbourg and June 30 of the same year. I began the final march on July 1, as the 4th Infantry Division and its 22nd Infantry Regiment turned their eyes east, toward Germany.

For me, getting to Gmund with my father was now almost secondary to learning from the men who would help to get him there.

Part Seven

FOOTSTEPS TO BAVARIA

"Incentive is not ordinarily part of an infantryman's life... Instead, the rifleman trudges into battle knowing that statistics are stacked against his survival. He fights without promise of either reward or relief. Behind every river, there's another hill—and behind that hill, another river. After weeks or months in the line, only a wound can offer him the comfort of safety, shelter, and a bed. Those who are left to fight, fight on, evading death but knowing that with each day of evasion they have exhausted one more chance for survival. Sooner or later, unless victory comes, the chase must end on the litter or in the grave."[109]

—**General Omar Bradley.** *A Soldier's Story*, 1951

Interrogating a German Officer, Lieutenant Sisson front right, October 1944

"F Company Rides, April 1945" [110]

30. Deadly Hedgerows

JULY 1 – AUGUST 16, 1944

The capture of Cherbourg meant Allied soldiers and supplies could now come ashore more safely at Utah and Omaha beaches. Replacement troops had to adjust to the realities of a brutal conflict and the exhaustion of their already war-hardened colleagues. Lt. George Wilson landed at Utah Beach and Sgt. James Cullen landed at Omaha Beach during the first week of July. Wilson noticed the terrible scars of war as he came ashore: burned out vehicles, twisted armament, and stern German prisoners of war sometimes vomiting from the task of reburying their dead comrades.[111]

Sgt. Cullen: Whatever the name, there were 343,648 "reinforcements" in the 1st Army in the ETO during the war…Just as the ammo was expended and replaced, so were the front-line infantrymen.[112]

July 1–6, 1944: With the port of Cherbourg secured, the Second Battalion headed south from Saint-Pierre-Eglise to train and reorganize near Gourbesville and Amfreville, then moved to Appeville as part of the division reserve.

JULY 7-12, 1944, BATTLE OF PÉRIERS

The town of Périers was the regiment's next strategic objective. On July 7, the Second Battalion advanced with the 22nd Infantry to a position three and a half miles west of the small but critically vulnerable seaport town of Carentan to reinforce the 83rd Division. They were to advance southeast toward Périers along a narrow ridge hemmed in by hedgerows and marshland on the Coutances–Saint-Lô Highway. It was a dangerous undertaking.

July 8–9, 1944: The 22nd Infantry Regiment's mission was to breach the enemy line south of Culot, with La Maugerie as their initial objective.

Chaplain Boice: The fighting began on the morning of July 8th and soon became one of the bitterest battles of the war…[As the troops slept uneasily the night of July 9th,] it was rumored throughout the command that the Germans had forty tanks in Periers which they intended to commit some place on the line…water seeped into foxholes and clothing was clammy.[113]

July 10, 1944: The leadership of the 22nd Infantry Regiment changed radically when a new commander introduced himself by field phone to the headquarters staff in the morning.

Col. Lanham: I am Colonel Charles T. Lanham. I have just assumed command of this regiment, and I want you to know that if you ever yield one foot of ground without my direct order, I will court-martial you.[114]

July 1–15, 1944: Hedgerow Battles

Chaplain Boice: It was a proper introduction to "Buck" Lanham. But the regiment began to fight with a skill, imagination and daring it would not have attempted before.[115]

That day, during the bloody battle for La Maugerie, Lt. Jim Beam, F Company C.O., was seriously injured when a mortar shell landed close to his feet. His right foot was almost severed, and his left foot badly broken.

Chaplain Boice: Lt. Beam calmly lit a cigarette and assisted Cpt. Humm, the Bn. Medical Officer, in completing the amputation of his foot. Having given himself a morphine surette, he refused to be evacuated until he had completed his orders, reorganizing his company to withstand the fierce German attack. Such were the men of the Twenty-Second Infantry.[116]

Dad said of Jim Beam,

```
His good humor and ability got us over
many rough spots. I don't believe I ever
saw him either grin or scowl at the wrong
time—except when he cleaned me out in
a particularly juicy poker game. I nev-
er did figure out his approach to English
girls. It was infallible.
```

Pvt. Laurel Pierce, one of my father's favorite platoon members and Jim Beam's close friend, was killed by the explosion that severely injured Beam. My father noted in his story that Beam was not told about his death until much later. F Company was paying too high a price.

Chaplain Boice: Private Hicks, who was then known as the "human tank-destroyer," got his third tank from the corner of a hedgerow behind a tree. He got three bazooka shots into the German tank at less than five yards. He was so close that the explosion scorched his face.[117]

July 11–12, 1944: On July 11, La Maugerie was taken.

Chaplain Boice: The attack on the 12th was to prove extremely costly in lives of both officers and enlisted men. The Germans were waiting and well prepared…The attack had only begun when Captain James B. Burnside, Second Battalion Executive Officer, was wounded, leaving the battalion with only seventeen officers. The fighting became fierce hand-to-hand combat. The rifle companies did not have over seventy effective fighting men. Though the number was small, individual courage and initiative was everywhere apparent. First Sergeant Kenyon, of G Company, gathered together fifteen men and took over a section of the front normally held by a platoon. He stated in a rather calm undisturbed manner, "I've taken over this part of the front and I'm going to hold it. You don't need to worry about it."[118]

Capt. Henley: (7/12) AT [Anti-Tank] platoon lost 17 men out of 30 in two days of fighting.

Boice tried to express what the men were experiencing. The battalion commander, Lt. Col. Earl W. Edwards, was inspecting the Second Battalion following fierce fighting when he came upon a medic who had been on the front lines every single hour since the battalion had landed on D-Day.

> When the Colonel approached, he was doing his best to aid a man whose skull had been laid open and who had sustained such a severe brain injury that he could not possibly live, and yet he would not die. The medic had done everything he knew how to do. When he saw the Colonel, his reserve broke and he cried, "Sir, he won't die. He ought to die, but he won't. I have done everything I can for him, but he won't die. Why won't he die?" The Colonel led the aid man away, giving instructions for much needed rest and care for the boy. This was war at its lowest, purest hell.[119]

Although my father never described Col. "Lum" Edwards in detail, he was one of the few ranking officers he called by first name. This story, and others, brought home his empathy, which no doubt helped make Edwards an unusually approachable leader of men.

July 13–15, 1944: By the morning of July 13, little was left of the proud fighting 22nd Infantry Regiment that had faced a determined enemy. The unit had suffered 1,375 casualties in less than a week of fighting. The advance to Périers was halted. By July 15, the regiment had moved to a training area near Carentan at Montmartin-en-Graignes.

Unaware of the recent bloody fighting, Lt. George Wilson reported for duty at the 22nd Infantry Regiment headquarters. His first introduction to the regiment was Colonel Lanham.

Lt. Wilson: A small, wiry man who looked as tough as he was gruff. He wasted no time in scaring the hell out of us. "As officers I expect you to lead our men. Men will follow a leader, and I expect my platoon leaders to be right up front…Use every skill you possess. If you survive your first battle, I'll promote you. Good luck."

Our guide was a corporal who was tired, hollow-eyed, and jittery. He acted like a cornered animal. Just watching his actions gave one the creeps as, bent low, he ducked and ran from one piece of cover to the next…Bodies of dead Germans were strewn along the way…The awful odor of death was increased by the hot July sun. Our guide told us our dead had already been moved, and we were grateful.

Headquarters was just a small ten-by-twelve tent set up under a tree near a hedgerow. Lieutenant Colonel Lum Edwards was in command. He greeted us briefly but made no speech...Lieutenant Piszarak and I were assigned to E Company and became very good friends and served together from July until November, when he was killed in action.[120]

The experience of the 22nd was repeated throughout Normandy: exceptional individual bravery, unacceptable loss of life, and little ability to advance against a well-entrenched and determined enemy defending a battlefield behind hundreds of hedgerows. Total casualties for the first five weeks of fighting stood at 3,439, more than the total strength of the 22nd Infantry Regiment on D-Day. The First Army had sustained approximately 40,000 casualties since Cherbourg; 90 percent were infantrymen. The Allies had to find a way to shift from the hand-to-hand fighting in order to expand their foothold and gain the room for maneuvers, airfields, and the increasing numbers of troops and supplies.[121]

July 16–24, 1944: On July 19, the rubble that was once the ancient city of Saint-Lô came under Allied control. The razing was the second in its history; the first was in 889 by Vikings. The city stood at a crossroads that was key to the Allied battle plan.

New tactics called for the full and coordinated use of air, artillery, infantry, and armor. Combat Command A, under Gen. Maurice Rose, was formed from the 2nd Armored Division and the 22nd Infantry Regiment, creating an armored infantry force. Its mission was to break through the German defenses south of Saint-Lô. General Patton and the Third Army would be poised behind them to thrust out of Normandy.

Training with the 2nd Armored, the men of the Second Battalion learned a new way of attacking that called for tanks to break into a field and spray the next hedgerow with their machine guns while the infantry walked or ran behind the tanks, using them as a shield. As they closed on the hedgerow, the infantry would run ahead and throw grenades over it. The tank would then plow through the hedgerow while the troops followed and captured any enemy soldiers.

Following the landing, tanks had been pushed upward as they advanced over hedgerows, exposing their vulnerable bellies. An ingenious soldier solved that problem. Tanks were now fitted with a scythe-like bumper made in part from the steel the Germans had used to create beach obstacles. This allowed them to plow forward through the hedges. Lieutenant Wilson recalled that the training went well, but that running in the heat of the summer behind huge vehicles that kicked up mountains of dust and spewed fumes was a challenge.[122]

```
Lt. Sisson, England: Dear Mother and Dad,
As I wrote you, pretty hurriedly I guess,
I was transferred to a new hospital...Have
run into one fellow from my battalion and
several from the regiment...Looks like at
least a month here. I'll probably have
seen half the officers in my regiment by
that time...Love, John
```

July 25, 1944: The troops assembled near Pont-Hébert in advance of the critical twenty-four-hour drive through the German line. Thousands of bombers passed directly overhead of the troops. The attack was to begin after the bombing stunned the enemy.

Chaplain Boice: Double Deucers counted as many as sixteen bomb craters per acre. Houses swayed as if made of some macabre papier-mâché. Mark V tanks were turned on their backs like toys, with the occupants killed or stunned by the concussion. German soldiers walked dazedly, unresisting, and unknowing, their eyes glassy and mouths gaping...[123]

Lt. Wilson: Staggering misfortune stepped in, however, with a cruel blow. One whole wing of bombers miscalculated and dropped its entire load on the front lines of one of our divisions. The losses in dead and wounded were over eight hundred, including the killing of Lieutenant General Leslie McNair. This tragedy and its confusion caused the postponement of the attack one more day.[124]

JULY 26–29, 1944: SAINT-LÔ BREAKOUT

Just after midnight on July 26, the regiment moved again to a forward assembly position north of Pont-Hébert. They attacked to the south at 9:30 A.M. in a single column, using the tactics they had learned. They would revert to attacking in two columns that afternoon.

During July 26–27, the 22nd charged forward west of Saint-Lô into the enemy. They fought through the town of Saint-Gilles, passed the haunting flames of buildings burning at night in Canisy, and finally halted at 2 A.M. outside Le Mesnil-Herman.

Lieutenant Wilson remembered a night filled with concentrated shelling. Fortunately, a number of the shells fired from the fearsome German 88 mm guns did not explode. He later learned the French Resistance was sabotaging 88 mm shells.

During the morning of July 28, the regiment consolidated

its position while patrols skirmished fiercely with the enemy. At the close of day, the Third Battalion was near Percy, the Second Battalion was just south of Le Mesnil-Herman, and the First Battalion occupied Moyon Villages.

July 29, 1944, Lt. Wilson: Our battalion was ordered to clear the Germans off a high ridge several miles long running parallel to the Villebaudon–Tessy-sur-Vire road. Other units attached to us, thus making a combat force, were a company of seventeen Sherman tanks, each with two .30-caliber machine guns and a 75mm gun. Supporting us were a platoon of tank destroyers...plus, artillery and mortars, and a cannon company with four 105 mm howitzers...By nightfall, nine of our seventeen tanks would be demolished, and the infantry would be almost wiped out.

Our ruination was the famous German 88...Its power was awesome. A direct hit did not bound off the sloping four-inch solid steel armor-plated front of a Sherman tank; it went clear through and out the back.

The calm hillside exploded into a full-scale battle...Suddenly we were caught between two fires—the Germans to our front and our own efficient artillery to the rear...Few things are as terrifying as the target area of an artillery barrage. You cannot think, cannot talk, and there is no place to go. You must fight your instincts to get up and run...

Tanks and tank destroyers were the exception. They had to move out of there...The withdrawing tanks thus could not see some of the men on the ground and the men, because of the overpowering din of the explosions, could not hear the tanks coming. Two of my men were crushed by the maneuvering tanks. I told myself they were already dead from the shells...Another of the men on the ground next to me was killed instantly by a mortar shell that landed on his back. His buddy and I were splattered with flesh and blood but were not touched by shrapnel.[125]

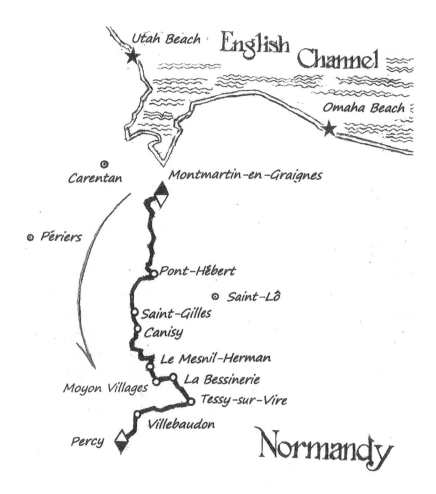

July 16–31, 1944: Saint-Lô Breakout

July 30–31, 1944: The next day, the battered Second Battalion was relieved. Though the cost had been high, they had accomplished a critical mission. General Patton and the Third Army had followed the Combat Team through the opening in the German line at Saint-Lô, and Allied troops were preparing to advance out of Normandy.

August 1–5, 1944, Capt. Henley: (8/1) Received order at 0215 to take Tessy Sur Vire…a helluva good tank battle took place…We lost 2—the jerry 3…Col. Teague's jeep ran over by Mark V tank. Colonel lost everything he had, except air mattress—"I'd commit the Bn. to save that."

Following the Third Battalion's attack and seizure of Tessy-sur-Vire, the rest of the 22nd Infantry Regiment would be released from the Combat Team and return to the control of the 4th Division. General Rose stated that he had never fought with finer infantry than the Twenty-Second. The 22nd Infantry Regiment had worked seamlessly with elements of the 2nd Armored Division. For their role in the operation, the regiment would receive a Presidential Unit Citation for extraordinary heroism and outstanding performance from July 26 to August 1, 1944.

On August 2, the 22nd Infantry Regiment assembled at Villebaudon and moved by truck convoy southwest to the town of Villedieu-les-Poêles to rejoin the division. On August 3, they began establishing roadblocks and preparing to attack Saint-Pois, which was captured on August 5.

August 6, 1944: From Saint-Pois, the Second Battalion cleared the high ground as far south as Chateau de Lingeard. They were ordered to form a defensive line with their regiment, beginning at the chateau when the Germans began a counterattack.

Capt. Henley: Jerry counterattack breaks thru near Mortain and gains success, but finally stopped. Air Corps and 3rd AR [armored division] trapped German column and annihilated all vehicles. Bn. alerted to assist but never moved out.

AUGUST 7-13, 1944, BATTLE OF MORTAIN

The German counterattack (Operation Lüttich) sought to break through the Allied lines behind Patton's tanks and cut the Allied supply lines. It was focused on the village of Mortain, which was defended by the American 30th Infantry Division. The 3rd Armored Division and the 22nd Infantry Regiment were assigned to the 30th Infantry Division during the battle. Sergeant Cullen, with the 3rd Armored Division, 36th Armored Infantry, fought just outside Mortain. Wilson and Cullen both noted fighting at a crossroads near the village. I wondered if they had crossed paths.

Sgt. Cullen: In our part of the battle, we could only see from hedgerow to hedgerow. It was a flat, narrow world and it became a noisy one…The rain of shells lasted for ten to fifteen minutes at a time. Then there would be a pause when we could hunt for casualties, go to the john either right in the hole or crouched close to the hedgerow. We would grab a bite to eat from our "K" ration or "C" ration cans and have a cigarette. Then we could hear someone yell "Incoming!" and the hellish noise and concussion would start again…We, all of us there in the line, kept our eyes squeezed shut and begged the Good Lord to stop the awful rain of steel.[126]

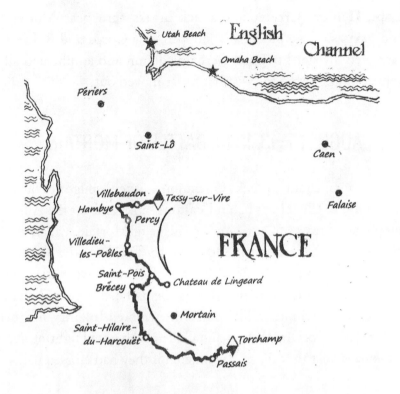

August 1–16, 1944: Mortain and the Falaise Pocket

August 10, 1944: The 22nd Infantry Regiment was ordered to leave the battle area and move rapidly south with the 4th Infantry Division to form a defensive line at the Varenne River near Passais. This would be their strategic role in the Falaise Pocket or Gap in which the Allies sought to encircle two of the primary German armies during August 12–21 following the failed counterattack at Mortain. The men dug in and awaited a battle, which did not occur.

August 14–16, 1944: Patrols continued to be conducted across the river, some working with the French underground. A small bridgehead was established at Torchamp. Otherwise, the men waited.

Capt. Henley: (8/14) If this is war, we like it. No artillery for 7 days now and this place is paradise.

Meanwhile, after the battle ended near Mortain, Cullen and his men broke out the calvados hidden in their jeep.

Sgt. Cullen: We didn't drink that much actually; we were all pretty young and none of us were drinkers. We were still learning the game…The battle was over, and we had been close to death, and we were high with adrenaline. We were still laughing as we dug shallow holes for the night. We took no chances and posted guards as usual…The coolness the squad had shown me when I joined them…seemed to have disappeared…By taking that fatal patrol [a dangerous two-man mission during the battle], they probably figured that I wasn't the type to avoid any duty that I might order them to do.[127]

Waiting or fighting, these young men were no longer untried citizen soldiers.

31. The Race Is On

AUGUST 17–SEPTEMBER 10, 1944

Gen. Bradley: September 1944 marked the month of the big Bust. But before the bust came the boom and until the Germans halted our tanks in the dragon's teeth of his Siegfried line, we bounded along on the headiest and most optimistic advance of the European war.[128]

Although the Allied attempt to encircle the Germans in the Falaise Pocket did not accomplish all its goals, it was a decisive battle in which major elements of two of Germany's army groups were all but destroyed. Allied troops moved rapidly into France. Starting August 17, the 22nd Infantry Regiment moved with the 4th Infantry Division from their position near Torchamp on the Varenne River to Carrouges then through Alençon and Chartres to just outside Paris on August 24. They were traveling in trucks, stopping only for fuel or to clear pockets of enemy soldiers.

August 17–24, 1944: During the Saint-Lô Breakout, Lieutenant Wilson had watched as one of his squads and the tank they were riding on were blown up, and the tragic death of a medic had moved his fellow soldiers to tears. He described the emotional turmoil that surrounds shooting another human being point blank, and what the infantry soldier experiences during an artillery battle.

Now, recalling the feeling of helplessness he had experienced during a recent bombing, he understood the terror German soldiers must have experienced in the face of devastating Allied bombing as they fled to the Seine after the failed German counteroffensive at Mortain. The carnage was terrible to see. "The dead Germans were literally stacked by the hundreds—in some places two and three feet deep. It was a real massacre. All the roads for miles were strewn with German corpses and littered with hundreds of smoking or burning tanks, trucks, and wagons."[129]

Capt. Henley: (8/18) Still counting our change—listening to good news of Patton's army heading for Paris—landing of American Army in Southern France.

Private Cassar was ahead of the 22nd Infantry Regiment, advancing with the Third Army's 7th Armored Division. They took Chartres on August 18.

Pvt. Cassar: (8/18) In Chartres it began, not in daylight but at night in convoy when snipers raised hell. At first all we could hear were our tanks' engines. Then their noise was drowned out by machine gun fire coming from the woods about a hundred yards away on each side of the road. Every time I heard the ping of a bullet hit the truck, I wanted to duck, but I had to concentrate on following the truck ahead of me. Adding to that was the sound of what I eventually could always identify as 75-millimeter shells. I was really in a war. I was scared! There was nowhere to go but forward, and I was driving a truck loaded with ammunition.

Celebrating Parisians greet the 22nd Infantry Regiment, August 5, 1944

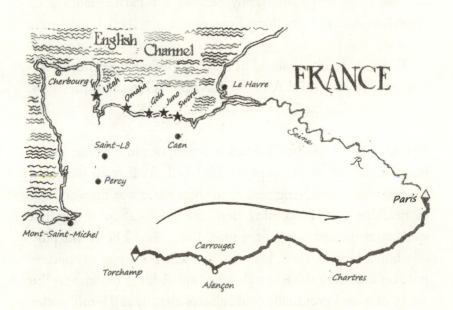

August 17–26, 1944: Paris Liberated

AUGUST 25, 1944, PARIS LIBERATED

Gen. Bradley: "To hell with prestige," I finally told Allen, "tell the 4th to slam in [to Paris] and take the liberation." Learning of these orders and fearing an affront to France, LeClerc's troopers [the French 2nd Armored Division] mounted their tanks and burned up their treads on the brick roads.[130]

The 12th Infantry Regiment moved by motor vehicle from their assembly area at 6 A.M. through Paris's southern suburbs from Longjumeau through Villejuif and entered Paris just after noon. They proceeded to the Hôtel de Ville, arriving at 1:15 P.M. after the French, who negotiated the terms of surrender.

The 12th mopped up scattered snipers in southeast Paris. The rest of the 4th Division (the 8th and 22nd Infantry Regiments) soon followed them. Enemy opposition in Paris was very light. The main obstacle to a rapid advance was the frenzied zeal of the populace itself. The reaction of the soldiers, whether officer or enlisted man, to the instantaneous party was about the same:

Sgt. Carlton Stauffer, Company G, Second Battalion, 12th Infantry Regiment: About 0800 hours on the morning of August 25, we began to move into the city of Paris...We all felt an exhilaration that would not be surpassed in the lives of any of us infantrymen. As we entered the Rue d'Italie, our tactical motor march became a huge victory parade, and our vehicles became covered with flowers. The pent-up emotions of four bitter years under the Nazi yoke suddenly burst into wild celebration, and the great French citizens made us feel that each of us was personally responsible for the liberation of these grateful people. We felt wonderful!

...At 0930 hours on the morning of August 26, Father Fries,

our regimental chaplain, held Mass in the famous Notre Dame Cathedral, the first mass said after the liberation. Joe Dailey and I attended. It was a strange sight for Notre Dame to see us doughboys sitting at Mass with our rifles and battle gear.[131]

Lt. Col. Ruggles: Bert Pokol was a very young soldier of Hungarian descent assigned as my jeep driver...Thanks to the 12th Infantry Regiment's liberation effort, the 22nd Infantry Regiment moved through Paris motorized. It was a start-and-stop move as the streets were jammed with celebrating French citizens hugging and kissing soldiers they could reach. Those of us riding in jeeps were really mobbed. Steel helmets were not designed for wear in this kind of an encounter. Pokol, that handsome devil (with helmet abandoned), was having a hard time driving. He turned to me during a short break in the assault from beautiful French women and said, "Sir, you will get some of this if you get rid of that cigar," then he added, "and the helmet, too." He was right.[132]

Capt. Tommy Harrison, HQ, Second Battalion, 22nd Infantry Regiment: The night before we were to take over and fight for the rest of Paris that was still in German hands, Colonel Lanham took about six of us to a café that Ernest Hemingway used to visit...Hemingway had left word that we would be there that night. The owner treated us like heroes and even opened a wine cellar he had hidden from the Germans...As you know, we all were very proud that we were there, "the day that Paris fell."[133]

Not everyone was cheering.

Chaplain Boice: Standing back from the crowd, under the awning of a sidewalk shop, stood a white-haired French woman. The lines of suffering and grief were deep upon her face, and she was old. No cheering here, and as the Americans rode by, she watched them unsmilingly. Clutched in her hand she held the Croix de Guerre. The ribbon was faded, and we suspected she remembered another war and a son who had given his life for France. The tears rolled unheeded down her cheeks…Wounds of the heart can be sharp and deep, and long of healing.[134]

August 26, 1944: The Germans fought a delaying action in Paris using small groups of infantry and tanks sheltered in buildings or behind walls. Armored action was limited to the shoot-and-run tactics. (After Action Report)

Lieutenant Wilson described an "emotion-ridden French kangaroo court." The mostly female defendants were being tried and convicted in minutes for collaborating with the Germans. The court was on a porch of a private home where a barber shaved the heads of the convicted. They were then lined up and forced to march through the streets, where they were jeered and "pelted with rotten eggs, tomatoes, and even bags of excrement" by the mob.[135]

August 27, 1944: And then, barely able to savor their triumph, the 22nd Infantry Regiment with the 4th Infantry Division was heading for Germany. The After Action Report for August reported total casualties since D-Day of 10,441 men or nearly 65 percent of the 4th Infantry Division's strength on D-Day. Chaplain Boice wrote that everyone believed it would be only a matter of a few weeks before the retreating German armies would be forced to surrender.[136]

Lt. Sisson, England: Dear Mother and Dad, Did you read about the 4th in Paris? How about that! When I hear about them, I really get an itch to get back. It's a funny sort of feeling that that's where you belong. If it weren't for the fact that I'd like to have another one of dad's Manhattans in the front room at 315 Gill Ave.—I think I'd be trying everything I could to get back…

August 27–September 10, 1944: On the Front Lines

August 28–September 10, 1944: The 4th Infantry Division advanced from Bois de Vincennes without encountering enemy front lines and secured a bridgehead across the Aisne River in Soissons on September 1. There had been intense fighting with the Germans in this part of France during World War I, and now the Allies were fighting again in the same places.

Capt. Henley: (9/1) Battalion coiled up for night. Rain set in again. Saw buzz bombs in flight. They really lit up whole sky and sound like 2-cylinder outboard motors skipping. Col. Lanham wants to win the war all by himself. Tommy Harrison still stuttering—smoking—and has the crud [any unidentified disease, from athlete's foot to intestinal problems].

September 3, 1944: Near the French-Belgian border a brutal battle had taken place. The troops were shocked to discover German horse-drawn equipment from World War I were used. Chaplain Boice noted the horses were groomed with their tails bobbed. Mortar, artillery, and air support had created terrible carnage. Men shot the desperately injured animals as they passed in trucks. It was one of the worst massacres Boice witnessed.[137]

September 5, 1944: As the 22nd Infantry crossed the Meuse River into Belgium at Fumay, Chaplain Boice reported two critical problems. Enemy resistance increased, as did Allied supply shortages. Gasoline was so scarce trucks moved one battalion at a time.

Private Cassar, advancing south of the 4th Infantry Division, was part of the liberation of Verdun on August 31. He drove through Soissons with German prisoners, four days after the 22nd passed through.

Pvt. Cassar: (9/5) The next day, I drove prisoners of war to a stockade north of us. As we went through Soissons, the people threw everything but bullets at the prisoners. I felt no pity for the Germans. War is brutal no matter what, but when humanity is eliminated, it reverts to barbarism. They fire at our medics and medical vehicles although they know, according to the Geneva Convention, medics have no guns with which to defend themselves. Conversely, we allow their medics to treat their wounded and then let them return to their lines.

There's something about war that could easily bring out the Attila in us that has to be contained. With the knowledge of what they've done to our guys that they've captured, I'm not certain of what I'd do, but all our German prisoners were delivered and will no doubt be better off than the ones that will have to face us in the next few days.

September 6–9, 1944: The 4th Infantry Division continued to advance to the east, meeting and overcoming pockets of enemy resistance, and maintaining contact with VII Corps on the left and the 5th Armored Division on the right. The continued shortage of gasoline and unfavorable terrain hampered the advance. They moved through Graide to Palliseul. (After Action Report)

Lieutenant Wilson recounted that in a small town in Belgium they arrived minutes too late to save a dozen people, mainly teenage boys, from being executed by the SS. Suspected of being an underground group, they were shot as a "lesson" to the community. The next day after he had left the village, he learned that the quiet couple who had taken him in for the night and served him breakfast with respect and kindness had lost their son to that "lesson." They had not mentioned it.[138]

September 10, 1944: The 22nd cleared the town of Houffalize with the enthusiastic assistance of the residents, who reported

German troop movements and helped clear roads. From there, they would advance into Germany. Lieutenant Wilson reported that the friendly warm smiles from civilians in Belgium disappeared as the troops approached Germany. There everyone stared straight ahead, fear frozen on their faces.

I wondered if the silence of the World War II generation that my friends and I had experienced also occurred in Europe. An American soldier returning home would not be reminded daily of what he had witnessed. In time, the psychological impact of the things that had happened overseas would begin to lessen. The urge for revenge that Private Cassar talked about would probably fade, too. This could never be true for the family of the slain boy in Belgium. They could revisit the place where he was shot, his gravesite, and their terror and grief at any time, any day.

I began to wake up from dreams at night thinking about these things.

32. Challenging the Siegfried Line

SEPTEMBER 11–NOVEMBER 6, 1944

The 22nd Infantry Regiment had reached the Siegfried Line, or West Wall. This was the German defensive system of bunkers, obstructions, and pillboxes that stretched for approximately 400 miles along the western border of the country.

Chaplain Boice: In certain sectors, the Siegfried defenses extended in depth as much as thirty-five miles from the frontier and consisted of a complicated maze of barbed wire and crisscrossed concrete and steel constructions, strewn on the soil like so many dragon's teeth; pillboxes and bunkers camouflaged into the surrounding landscape; fortress-like, armor-plated dugouts invisible to the human eye...[139]

Supplies had now become a truly serious problem for all the troops on the front lines, but the Americans moved resolutely forward.

Sgt. Cullen: The petrol supply line extended all the way back to Normandy, and we were out at the far end. Rumors then circulated that some of Patton's people had raided one of the 3rd Armored Division's dumps at gunpoint and had stolen several truckloads of gas. Patton supposedly clapped and laughed when he was told about it. In our minds, we then began to lump him in with the Krauts.[140]

SEPTEMBER 11, 1944, INTO GERMANY FIRST:

Chaplain Boice wrote proudly about the 22nd Infantry Regiment's achievement. "Late that night a strong reconnaissance patrol was organized with the mission of crossing the German frontier. The patrol was charged, in addition to obtaining enemy information, to return with a jar of German soil to be sent to the President of the United States.

When the patrol from the 22nd Infantry Regiment's Third Battalion reached the Our River, they realized they had to continue on foot so the two commanders, Lt. Manning and Lt. Shugart, flipped a coin to see who would lead the patrol. "Lt. C. M. Shugart, the I and R platoon leader, won the toss and led the foot patrol into Germany." They entered near Hemmeres at 2130 hours in what would be credited as the first Allied unit to cross onto German soil during World War II.[141]

Simultaneously, Lieutenant Wilson and the Second Battalion advanced south of the Third Battalion through Steffeshausen and Auel. He led a small patrol that crossed the German border into Winterspelt and north to Bleialf. From there, on the road to Sellerich, Wilson set out on foot and saw the Siegfried Line defenses. He was called back and joined E Company south of Bleialf.

September 12–13, 1944: Intermediate objectives were secured in the vicinity of Saint Vith. The battalions moved to the Bleialf area to prepare for the attack on the Siegfried Line. (After Action Report)

Tanks couldn't cross these, so they drove down the roads. September.

Armored support for Siegfried attacks. The 22nd penetrated the Siegfried line on September 14, 1944.

Siegfried Line defenses near Elcherath, September 14, 1944 (Dragon's Teeth on left.)

September 11–12, 1944: Into Germany First Detail

Capt. Henley: (9/12) Jumped off at 1130 AM for action. German border. Crossed river into actual German border at 1730.

Forded river, as bridge was blown. Entered Hemmeres and Elcherath — my what a cold reception the civilians gave us. No more liberating towns — it's conquering now.

Ernest Hemingway had joined Colonel Lanham's proud 22nd Infantry Regiment, one of his favorite infantry units, to report on its progress. It is said he gave a victory dinner in the Hemmeres post office for the officers of the 22nd. Henley's diary does not mention the party.

Col. Lanham: The food was excellent, the wine plentiful, the comradeship close and warm. All of us were as heady with the taste of victory as we were with the wine. It was a night to put aside the thought of the great West Wall against which we would throw ourselves within the next forty-eight hours. We laughed and drank and told horrendous stories about each other. We all seemed for the moment like minor gods, and Hemingway, presiding at the head of the table, might have been a fatherly Mars delighting in the happiness of his brood. It was the last happy day.[142]

North of the 22nd Infantry Regiment, the 33rd Armored Regiment commanded by Lt. Col. Lovelady advanced toward the German border with the 3rd Armored Division. Sergeant Cullen noted an increase in enemy resistance and the growing lack of cheering residents. Soon there were only empty villages with white sheets in the windows. On September 12, they took the German town of Roetgen. There was no famous novelist reporting on his regiment, but for them, too, that day was the end of their jubilation.

Sgt. Cullen: All us tankers and infantry alike, jumped down at the first stop and peed on German soil…There were concrete dragon's teeth on both sides of the road, and a steel gate in the middle. We were not going to get through that mess.[143]

The next day, during a ferocious attack just outside of Roetgen, his words proved prophetic:

Sgt. Cullen: As I raised myself from the ground, there was the ringing crash of a shell hitting my half-track. At the same instant I felt a blow to my chest, and I was knocked back to the ground. I had been hit! I pulled my shirt open at the neck and looked down and saw blood pouring from my chest—right over my heart. There was no great pain, only a feeling of shock and surprise…Was this the end? Was I going to die in this lousy German field? How would Mom, Dad, Billy, and Martin get the news?…There was a lot of blood, Better get it stopped!…Savage and I staggered down the road toward the rear…The aid station was in a ditch by the road and seven or eight of us were stuffed into the meat wagon and were carried farther back to a clearing station.[144]

September 14–October 3, 1944, Lt. Wilson: Colonel Lanham also had an uncompromisingly aggressive nature. He believed the best way to end the war quickly and save lives was to attack and attack…He wanted his regiment to be the first Americans through the line, as they'd already been first across the border into Germany. Plans were to attack east from the vicinity of Buchet.[145]

Capt. Henley: (9/14) Attack Siegfried line at 1000 and penetrated it. 3rd Bn turned right and 1st Bn turned left. 2nd Bn in reserve... The German West Wall = pill boxes galore and tank traps—dragon teeth mines. Lived in jerry pillboxes in West Wall for 3 days. Our Bn covering 2000 yd front—something for the tactics expert to write about. Casualties heavy—88 the 1st day—58 the 2nd day—approx. 50 the next day. Fought pillbox after pillbox—took lots of prisoners (SS troopers).

September 12–November 6, 1944: The Siegfried Line

The 22nd would advance deep into the West Wall, halting outside Brandscheid. Colonel Lanham's request to continue the attack east toward the Rhine River was denied by division headquarters. The regiment's attacks would be confined to the immediate area for the next three weeks.

Lieutenant Wilson wrote that during this period his platoon witnessed a failed attack by another infantry regiment that charged into German artillery and suffered terrible casualties. The next morning, he had to call upon his own men to launch an attack. When one of his best soldiers started to climb out of his foxhole he collapsed in a fit of convulsions. Wilson had noted, "How do you forget such carnage while trying to rest up to repeat it?"[146]

September 17, 1944, Capt. Henley: Still in position in the German west wall about 2 miles north of Brandscheid. Ordered to continue mopping up of pillboxes in immediate vicinity and continue on to clean out boxes to Brandscheid. Moved on town itself with tanks and TDs and was ordered to withdraw and move at once to help 1st Bn who were cut off by counterattack. Jerry artillery and mortar fire heavy—enemy sniper active. We called on Corps artillery for counter battery, the artillery dueled all day and night.

September 18, 1944: The 4th Infantry Division would press on and make strong inroads but could not take control of Brandscheid. Casualties and the lack of supplies took their toll, and on September 18 the 4th was ordered to secure its positions at Buchet and cease further attacks on Brandscheid.

```
Lt. Sisson, England: Dear Mother and Dad,
Well, it looks like we're off to the
wars again...The way things are going over
there, I'll have to run like the devil to
```

> make the lines before they cease firing.
> I'm really hoping to get back to the out-
> fit. Have run into several fellows around
> here, in various stages of recuperation,
> and we've got a reunion all planned in
> Berlin…Lots of love, John

Illustrator John Groth was with the American press who traveled north from Luxembourg to report on the 4th Infantry Division which had penetrated the German line. He described Hemingway's farmhouse, which became the center of press activity—and of danger. German patrols were passing within 50 yards of them at night.

The 22nd Infantry Regiment's Lt. Col. Teague, the commanding officer of the Third Battalion, escorted Groth into the battlefield to observe the action and to draw what he witnessed.

John Groth: The battalion had captured thirty-one pillboxes, and in nearly every case there were no live Germans. We crawled to another pillbox that had been captured the day before…It had been manned by sixteen men; there were sixteen dead men when it was opened. There were several Russians and two Hungarians. There was only one SS man among them. In their pockets were the Hitler orders to die at their posts; and in the pocket of the SS man's blouse, another order: "If the Cossacks don't fight, shoot them."[147]

Capt. Henley: Still in position and ordered to be ready to move out on short notice with tanks and TDs. Lt. Col. John Dowdy killed by jerry artillery at Sellerich. Returned to Brandscheid to cover up pillboxes with bulldozer. Rode in tank as assistant gunner. Set 6 houses on fire—blew up 4 haystacks and put long range tank fire on herd of cows and a flock of chickens.

I decided this last entry described Henley's way of dealing with the mayhem and tragedy all around him. Although the Combat Team was not advancing forward, enemy counterattacks were numerous (and successfully repulsed), patrols were aggressive, and contact with the enemy constant.

September 20, 1944, Capt. Henley: Still here counting our change. Jerry still throwing artillery and mortar. Major Latimer relieved of Bn and I took command of it. Capt. Tommie Harrison came over as Ex. Officer—Capt. Eggleston S-3. The surprise of my life when Col. Lanham called on the phone and said, "You are [Third] Bn CO—start functioning."

October 4–23, 1944: Chaplain Boice wrote that after just under three weeks of fighting several miles inside the Siegfried Line, the regiment was ordered on October 4 to move north to Honsfeld, Belgium. The 22nd Infantry Regiment would engage in combat for the month. The weather turned enemy with drizzling rain and penetrating cold. Patrols remained active around the towns of Miescheid and Udenbreth. Showers, movies, Red Cross facilities, and USO shows were set up in Mürringen to provide respite for the troops.[148]

John Groth: Foxholes were having to be bailed of ice water…Winter clothing had not come up to the front yet…I stopped to draw several GIs who were digging a new set of foxholes. Blanketed Private Robert Bovee, a nineteen-year-old from Pierpont Manor, New York looked up at me from the mud of a hole he was digging to tell me that this was the sixth foxhole he had dug that day—and the twenty-fifth since the beachhead. Private Doyle E. Stockstill, from Bogalusa, Louisiana, who was digging just beyond Bovee hollered over, "By God, when I get Hitler, I'm going to make him fill up every foxhole I've dug!"[149]

October 6, 1944: While my father's battalion battled through France, Belgium, and into Germany, he made his long trip back to rejoin them. Lieutenant Sisson was now one of the thousands of replacements constantly arriving to replenish the front lines. He began his journey from England on September 21, 1944. During the English Channel crossing, he met French resistance fighters and citizens returning home. Landing at Omaha Beach, he traveled by truck on the Red Ball Highway to Etampes just outside of Paris. From here he would be transported by train in 40 & 8 livestock cars through Paris, on to Arlon, Belgium, then north to Liege.

Amid all the chaos, two young men from Galion, Ohio whose mothers were good friends ran into each other while heading back to the front lines. They had both been seriously injured on June 10, Bridgeman with L Company, Third Battalion. They would now return together.

Lt. Bridgeman, Belgium: Dear Mother, I'm sitting on my steel helmet and writing on a ration box so don't expect too much legibility. John Sisson is here with me, I met him at the last depot, and we moved up here on the same shipment. He is fine and we have done a lot of talking about our trials and tribulations. We figure the people in London and Galion will enjoy this meeting greatly. We are back up where the living is primitive, mud and pup tents are admirable conditions for misery. Oh, for that hospital bed...Don't worry, I am in the best of health. All love, George[150]

I wondered how closely the two friends from Ohio were following the path their 22nd Infantry Regiment had forged just months earlier while fighting the German army.

October 10, 1944: Lieutenant Wilson and his platoon faced a new menace while on patrol. Hidden land mines now threatened them, particularly at night. He wrote that they were more terrifying than anything he had yet experienced. He was doing well, however, and wrote that his highly respected friend Capt. Arthur Newcomb, who had been on the battalion staff since shortly after landing with the regiment on D-Day, had been reassigned to the command of E Company.

October 12, 1944, Lt. Wilson: All my patrols of the past weeks had apparently impressed someone. Lieutenant Colonel Walker transferred me to battalion headquarters and promoted me to the job of Battalion Intelligence Officer (S-2). I was now responsible for getting all possible information about the enemy...Finding out enemy strengths and weaknesses, his gun emplacements, mine fields, recent changes in his defenses...it was not a desk job. We had to dig up this intelligence ourselves, and the only firsthand source was patrolling.[151]

October 15, 1944, Lt. Sisson: George and I got a ride up to headquarters and were received like a couple of long-lost sons. We had lunch with Col. Lanham and Lum Edwards.

OCTOBER 24–NOVEMBER 6, 1944, KRINKELT, BELGIUM:

Through the remainder of October and until November 5, the regiment occupied a position in the Krinkelt area that paralleled the Siegfried Line fortifications. This period was marked by sharp patrol clashes. There was considerable harassing artillery and mortar fire. The battalions, whenever in reserve, received training in demolition and mine removal and underwent physical conditioning. During the first days of November, part of the regiment was sent on pass to Paris and to the Corps Recreation Center at Eupen, Belgium.[152]

```
Lt. Sisson, (10/27): I went back to the
Second Battalion and went to E Company
for a day or two. Newcomb and Tolles just
had their whiskey ration, and we drank it
up in a dugout one night. We made a short
move to Krinkelt, where I was made S-2
under Glenn Walker. Rigged up electricity
with the aid of some Krauts and listened
to a US football game broadcast...Jer-
ry Claing had been fiddling with a ri-
fle grenade and fired it with a live round
in the chamber shortly before. Broke his
leg nicely, and Phillipi has F Co. now.
Newcomb had E Co. with Tolles as Exec.
Things were sure a helluva lot different
from the peninsula. At F Co. I found Pol-
lock now 1st Sgt. and in pitiful shape.
Got my first look at buzz bombs here. They
```

> were going about 200 feet over our house, headed for somewhere.

Lieutenant Wilson suddenly lost his position "as an S-2, as part of the elite," when Col. Walker informed him his combat experience was desperately needed in F Company. He thanked him for his excellent service and said he was very sorry. Wilson wrote, "I hated to start all over again, but I tried hard not to let it show."[153]

I thought about the amazing coincidence that I had found my father's and Wilson's correspondence about this incident forty-six years after it happened. I said a silent prayer of thanks that George Wilson's return to combat as a result of my father taking his spot at headquarters did not result in Wilson's death on the front lines. I was also glad he explained what the job of S-2 entailed. My father never got around to it.

Ernest Hemingway returned to the front with the 22nd Infantry Regiment several more times during the war and partially based his protagonist in *Across the River and Into the Trees* on Buck Lanham. The book was a commercial success but received negative reviews. In 1950, a review noted "there is hardly any aesthetic distance between the author and Colonel Richard Cantwell, the hero of the novel."[154] Hemingway never forgot the 22nd Infantry Regiment. He sent a turkey each year to their reunions after the war was over—the *Hemingway Turkey* is still a tradition celebrated by the 22nd Infantry Regiment Society.

Sergeant Cullen, though he had been severely wounded, would return to active duty. Unlike my father, he fought on the front lines again. When Cullen wrote his memoir, he ended by honoring the courage and fortitude of the soldiers and civilians he met during those difficult times.

More than sixty years later, a message sent to Col. Lovelady sums up the spirit of American soldiers and civilians over the ages, when faced with terrible odds.

When we crossed the German border, General Collins sent congratulations to Col. Lovelady. "Tell Lovelady he's famous. Now tell him to keep going."[155]

The 22nd Infantry Regiment had fought almost nonstop for five months, securing nearly 500 miles of enemy occupied Europe.

Chaplain Boice: They were in top peak; it was a fighting machine trained to an efficiency not matched at any other time during the war. It was an aggressive, battle-scarred, confidently experienced regiment bent on the destruction of the enemy.

The Landing at Utah Beach, the St. Lo Breakthrough, the Battle of Paris, the initial penetration of the Siegfried Line all stood as major achievements of the Twenty-Second Infantry…The regiment was poised for battle; it could not know the shattering hell of the eighteen days ahead.[156]

33. Forest of Death

NOVEMBER 7 – DECEMBER 5, 1944

The 22nd Infantry Regiment's battle in the Hürtgen Forest has been faithfully documented in *Hell in the Hürtgen Forest: Ordeal and Triumph of an American Infantry Regiment* by Robert Sterling Rush. It is an important book to read for anyone wishing to fully understand the 22nd Infantry's role in this battle called by most 4ID (4th Infantry Division) vets who were there the worst of WWII.

Gen. Omar Bradley: In any counterattack across the Plains of Cologne, von Rundstedt [the German Commander in Chief] could count upon a powerful weapon in the Roer River. As long as he held the huge Roer dams [located near Hasenfeld] containing the headwaters of that river, he could unleash a flash flood that would sweep away our bridges … Clearly, we dared not venture beyond the Roer until first we had captured or destroyed those dams.

… Not until three months later did we finally take those dams and thus secure the Roer for a safe crossing. Had we secured them in early November and pushed across the Roer, the enemy would never have dared counterattack us in the Ardennes.[157]

To capture the dams, the First Army had to first secure the Hürtgen Forest. The 28th, 8th, and 83rd Infantry Divisions had already been sent on this mission but had been beaten back. Now, it was the 4th Infantry Division's turn. More than 24,000 Americans were killed, missing, captured, or wounded during the fighting in the Hürtgen Forest, and another 9,000 succumbed to the wet and cold with trench foot and respiratory diseases. Many have questioned whether the battle should have been waged at all.[158]

Cpl. George Morgan, First Battalion, 22nd Infantry Regiment: Show me a man who went through the Battle of Hürtgen Forest and who says he never had a feeling of fear, and I'll show you a liar… You can't get protection. You can't see. You can't get fields of fire. The trees are slashed like a scythe by artillery. Everything is tangled. You can scarcely walk. Everybody is cold and wet, and the mixture of cold rain and sleet keeps falling. Then we attack again and soon there is only a handful of old men left… You can't get all of the dead because you can't find them, and they stay there to remind the guys advancing as to what might hit them.[159]

November 7–10, 1944: Division Headquarters established its new command post in Zweifall on November 7. Earlier that morning a quartering party of officers and enlisted men had arrived there.

```
Lt. Sisson: Around the first of November,
I was briefed on the coming operation,
which was to take Duren via the Hurt-
gen Forest. Looked nice on paper. Several
days later, I was on the advance party to
the Hurtgen. Story was that Krauts were
about a mile or less in front of us, and
```

```
that we could make reconnaissance. Super
hush-hush...Hitler probably knew more about
us than we did.
```

November 7–December 5, 1944: The Hürtgen Forest

When the blacked-out trucks carrying the troops finally arrived at their new post, Lieutenant Wilson wrote that Berlin Sally, the English language propaganda broadcast from Berlin, welcomed the 4th Infantry Division to the Hürtgen Forest.

```
Lt. Sisson: After a few nights in dug-
outs or pill boxes, the rest of the Bn.
```

came up and we put them into position. A beautiful night it was. Cold, snowy, sleet, freezing, foggy. Wait. Wait. Hours of waiting. Finally at dawn our Bn. came trudging along. I thought I had heard of artillery before. But when I saw patch after patch 200- or 300-yards square with trees completely blown out, and sometimes a half mile with all the treetops blown off, I began to reconsider.

November 11–15, 1944: The dark pine forest blocked the limited sunlight, so the men were in darkness from about 4 P.M. until almost 8 A.M. The dangers were intensified by the dense trees and the lack of roads. Medical aid men found it necessary to carry the injured on litters up to two miles over rough terrain and through extensive mine fields. The casualty rate among aid men and litter bearers was extremely high and replacements were difficult to obtain. (After Action Report)

November 16–December 5, 1944, Battle of the Hürtgen Forest: In eighteen days, the 22nd Infantry Combat Team moved 7,500 yards—less than 7 miles—to take Grosshau, which was twenty miles north of the high ground at Schmidt near the Roer River Dams. "Casualties within the rifle companies during this eighteen-day battle reached a staggering average of 138 percent of their original strength, including replacements." Of the soldiers who went into battle on November 16, 91 percent of enlisted men and 93 percent of officers were casualties by December 4. In a day, companies were reduced from 174 soldiers to as few as 87. Company commanders had a 300 percent casualty rate.[160]

Col. Edwards: I still flinch at the memory of those young kids being led up the dark roads at night to their units, with many being killed or wounded on the way.[161]

November 16, 1944, Capt. Henley: D Day, H Hour, 12:45

Chaplain Boice: As the account of the Hurtgen Forest is uncovered, one must keep in mind the horrible surroundings in order to appreciate the gallant struggle put forth by the determined infantry units...These heroic men fought continuously within fifty yards of the enemy, often with actual physical contact and with sure death only seconds away. These men ate the issue rations when they were frozen, muddy, and stale. Fires were unheard of. These men lived day and night in the bloody slime to be found only in the Hurtgen Forest. It was not an uncommon sight to see a dead soldier with the pit of his stomach ripped open...or to hear the wounded scream in terrifying pain with their arms or legs completely blown off by an enemy shell.[162]

```
Lt. Sisson: At 1245, we pushed off.
Climbed the damndest hill I ever saw,
then down again to tell Walker [Second
Battalion CO] the assault companies were
on top. No fight yet. The companies pushed
on. Bn went up the hill and set up on top
of the hill. Then we had it. Mortars for
about five minutes. I looked at my hand
and saw a little hole in my glove. Broth-
er, if there will ever be immunity to
fire, Sisson won't have it. Patched it up
and it didn't seem bad. We dug in on the
```

reverse slope that night after firing and being fired upon all afternoon. The krauts thought they were in these woods to stay. Every fifteen minutes we got another barrage. Started getting casualties.

November 17, 1944, Capt. Henley: All hell broke loose. Jerry artillery and mortars cutting us to pieces. Major Drake killed—Capt. Henley took over [First] Bn and fought with the men until they had secured objective and turned it over to Major Goforth who came up. Jerry artillery and mortars are still cutting us to pieces. Casualties high—between 70 and 100 men hit. Evacuation is extremely difficult. Had to hand carry the wounded about 1 mile. All supplies hand carried over extremely hard conditions.

Lt. Sisson: In the morning I made a reconnaissance to contact F Company, got about a half mile and found one squad trying to dig in. When I asked them where to Company CO they pointed ahead. Then I said, "Where are the krauts?" A guy poked his carbine around a stump, sighted something about a hundred yards away in the dense undergrowth and fired a few shots. "There," he said, "I think." Oh it was great in the Hurtgen. The whole First Battalion staff was knocked out at one time that day. Walker finally sent me back to have the hand dressed.

November 18, 1944: The next morning, the rest of the Second Battalion headquarters staff were injured and evacuated. Major Blazzard, who had landed at D-Day with the 22nd, took over the Second Battalion and advanced with them to where the First Battalion was embattled.

November 19–26, 1944, Chaplain Boice: Up front in the half light of the forest, in the frozen slush and deep silence punctuated by the staccato of the German burp gun fire and, answered by our M1 Rifle and light machine gun fire, the fight went on. Exhausted men swore softly and died or cried and collapsed as was their want, but the 22nd Infantry Regiment Combat Team of the Fourth U.S. Infantry Division fought and slugged and pushed and dragged its way eastward yard by yard, 500 yards a day, 1000 yards a day, 850 yards a day, 85 casualties a day, 143 casualties a day, leaving a trail of broken men, American and German side by side, to mark its progress.[163]

Capt. Billy Carter: I was Services Company commander. The HQ company commander was killed, and I was assigned as Headquarters CO ... The last available men from our regiment (I think about 135 cooks, clerks, and truck drivers) were assembled, and we were assigned a sector of the front line with me in charge.[164]

Capt. Henley: (11/21) Another close call. Lt. Hall hit standing right beside me in the ribs. I applied first aid and went for aid. These 5 days so far (Nov. 16-21) the roughest and bloodiest I have seen. It was a nightmare.

Pvt. Patrick O'Dea, Company F, Second Battalion, 22nd In-

fantry Regiment: Leaving the road, we passed stretcher bearers carrying a severely wounded man back to the aid station. Slipping in the mud, we went over the crest and started down the other side of the hill. We followed no trail, only our leader knew where we were going in the grim forest...It was on the hike that I saw my first dead man...It was extremely rough climbing, and before long the sweat poured off me, in spite of the cold...then it happened. There was a sudden short high-pitched scream, and a shell exploded nearby with a deafening roar. I was taken completely by surprise...Instinctively I hit the ground...another explosion shook the earth, then another, and then another. Our officer jumped up and said, "Let's get out of here." With a start, I realized that we had reached our goal. Here was Company F. We were standing on the front lines...[165]

At one point during the fighting, Lieutenant Wilson went looking for his medic and found him trapped in a foxhole under the bodies of a young replacement lieutenant who had been killed and Sergeant Servat, my dad's BAR man on D-Day. Servat was badly wounded but still showing his remarkable resilience. Once out of the foxhole, the medic began tending to the sergeant's wounds.[166]

November 27, 1944, Capt. Henley: B Company given same mission and after losing 70 men, took objective and dug in for night in edge of Grosshau. The most jerry artillery and mortar I have ever seen or heard of.

November 29, 1944: The Second Battalion attacked Grosshau at noon. The attacking force consisted of Companies E and F with armored support. Immediately upon jumping off, Company

F received a counterattack, which it repulsed. Company E advanced slowly under heavy German machine gun fire. Resistance in the houses and cellars finally ended around 7 p.m.

Lt. Wilson: I had reached the limit of my physical and emotional endurance. The barrage abruptly ended, and a problem with my radioman…snapped me right out of my morbid thoughts. He was crying again, this time with reason, and he begged me to send him to the rear. It wasn't the best time to bother me, and I couldn't take it from him. I turned on him angrily and pointed my rifle at his chest, saying that if I heard one more word out of him, I'd shoot. He stopped bawling instantly. [167]

Pfc. Benny Barrow gives a comrade a hand up during the bitter fighting in the Hürtgen Forest on November 18, 1944.

22nd Infantry Regiment in Grosshau, Germany, November 30, 1944

November 30–December 3, 1944: On Dec. 1, Henley reported the CP was in a cellar in Grosshau. Having secured the objective, they were ordered to advance on the village of Gey. The brutality of the next three days of battle on the Grosshau-Gey road was hard to fathom. The village was on the edge of the forest and the troops were attempting to drive the enemy from their entrenched position there.

Capt. Henley: (11/30) Sniper firing broke out in Grashau again…3rd Bn to move forward to phase line in edge of woods south of Gey.

Lt. Bridgeman: (11/30) The maneuver was very tricky. It is remarkable that only once did one of the tanks [attacking on both sides abreast of the Third Battalion infantry] get a little behind and accidentally shot four of our men.[168]

The next day the casualties were horrific.

Lt. Wilson: (12/1) We had lost all the medics, all the noncoms, three of the four company officers, and the artillery observer. And we had lost ninety riflemen. This was, and still is, the most terrible day of my life...The loss of one man is deeply depressing; the loss of ninety is just overwhelming. In fact, I was overwhelmed by the courage of these men...Emotion was all most of us had left.[169]

Capt. Faulkner: (12/1) Found F, Lt. Wilson and 15 men left out of 150.[170]

Writing from a safe vantage point behind the lines at the medical center, Lt. John Sisson's letters home were less lighthearted now. The banter and requests for items disappeared.

```
Lt. Sisson, (12/2): The Belgians are very
friendly, except that they think we're
taking it too easy on the German prison-
ers we have. They seem to have developed
some very hardheaded ideas about the al-
leged supermen. Have a few of my own, as
a matter of fact.
```

Wesley Trindal, F Company, Second Battalion survived the November 27 attack on Grosshau in which the nine other men in his patrol were killed. Lying unconscious in the trenches after the battle, he was taken for dead by the Germans. He finally made his way to a medic station in the now American-held Grosshau with the aid of two German soldiers. They had saved his life so he could bring them in as his prisoners. He kept his sanity by looking at the pictures in his wallet and focusing on memories of home and those he loved—alone, and then with the two Germans. From the medic station, he could see the ground they had advanced across when they attacked the village:

Pfc. Trindal: We were fodder for the German weapons. What targets we made! The German riflemen, machine gunners, tankers, and artillery crews had a clear field of fire on all of us as we advanced. No wonder we had such losses![171]

When the command to relieve the 22nd was formally received on December 3, Col. Lanham knew there was virtually "no more fight" left in his regiment. He did not let Captain Henley leave with the advance party to prepare the new headquarters, he considered him a "symbol of good luck to the Regiment" and there were still 20 hours before December 4 saw the regiment finally heading south.[172] The colonel was wise. One last German counterattack that next morning nearly succeeded.

Dec. 4–5, 1944: The 22nd Infantry Regiment completed relief and moved to an assembly area. They passed through Zweifall enroute to the vicinity of the city of Luxemburg. (After Action Report)

Chaplain Boice: CT 22 had completed its battle through the Hürtgen Forest. In eighteen days of the campaign the combat team suffered 2,575 enlisted men and 103 officer casualties … The Germans themselves could not believe that the Americans could continue to fight when they well knew the extent of the casualties. Thousands of men had passed through the Regiment in 18 brief days. Hundreds of men had lost their lives for a patch of woods and the heap of rubble that was Grosshau. Perhaps in the final analysis, the sacrifice demanded in Hurtgen will be deemed worthwhile; we wouldn't know about that. We were only the men who fought the battles...[173]

Col. Lanham, 9 December 1944:

To The Officers and Men of Combat Team 22:

...From the night of 16 November until our relief on 3 December, our Combat Team was the easternmost unit in the entire army and therefore the greatest threat to the German Reich. To stem our attack, units were withdrawn as far south as the Schnee-Eiffel sector and as far north as Merode and flung against us...There are no words to describe my pride in you or my confidence in you. I can only repeat what has been said to me again and again by those who know your record and who have seen you fight: "You are one of the greatest fighting teams in all American history."[174]

```
Lt. Sisson: After ten days I was released
at Liege. On the same day, Col. Bryant,
G-1, started pumping me about my records
with the Division. Ended up saying, "Son,
how would you like to spend Christmas at
home?" Two wounds and ten days hospital-
ization each, a 30-day temporary duty in
the U.S.A. was ordered by Eisenhower.
```

In his story, my dad had struggled with his loyalty to his battalion and the "minor" wound that took him from the front lines until he returned for the second time to witness the injuries sustained by the rest of the Second Battalion staff. On December 9, he would gratefully head to Paris and then home. During his leave, he would type the narrative his family found the day after he died.

John Groth was ordered home from the war about the same time. Apparently, as far as the news media was concerned, the war was over—Germany was licked. He too struggled with leaving what he knew to still be a VERY active war. He wrote about the reactions of the young men on the front lines when confronted with this news from the home front.

John Groth: One kid suggested: "Why don't they dig a string of foxholes in the wooded hilly sections of city parks, put ten representative civilians in 'em, turn on the rain that lasts all day, every day, and fire mortar shells at 'em. When the mortar shells get close, make 'em get out and dig another foxhole. When the mortars find that one, make 'em do it again. Make 'em feel like cockroaches dodging a boot. After a few days of this bein' wet and standin' in ice water with no hot food and no letter from home and not even a magazine, ask 'em if it's all quiet on the Western Front and the war's over." [175]

These same young men did not begrudge Groth his good fortune. They told him, "Any guy that can go home and doesn't want to is nuts." [176]

From what I had read, the better you were at combat, the more combat time you were given as a reward—until you were too injured to fight, or lost your mind, or your life. For too many riflemen, leaving the front line alive and intact, as my dad did, was always just out of reach, beyond the next advance into enemy fire and looming death or disfigurement. I had not understood what a soldier woke up every morning to face. And, if he was lucky, he went into his foxhole every night knowing what he would face again the next day. The boys who had come home from Vietnam, the young men I had shown so little respect for, had faced similar dangers.

Many years later, my father learned that his—and his granddaughter Kathryn's—favorite soldier, Sgt. Alfred C. Mead, died leading an F Company squad in the Hürtgen Forest. I cannot imagine how dad felt when he found out. When he wrote Al's name with the names of his other fallen comrades from his original platoon in his notes, he indicated that he died November 1945. In that simple mistake, I sensed my father's pain. I wondered if he knew that the man who saved his life had died in that cold unforgiving forest on November 17, 1944, the same day he was evacuated for the hand injury that would send him back to the States on leave. I hope not. Although so many had suffered far worse, I hated to think that my father had to live with knowing that.

Chaplain Boice: The wards with shattered and missing legs and arms were bad, but the hospitals with vacant and missing minds were worse.[177]

34. No Bulge Here: Luxembourg Defended

DECEMBER 6-31, 1944

In his history of the 4th Infantry Division, Robert Babcock observed that when we hear about the Battle of the Bulge, we most likely think of the 101st Airborne and their stand at Bastogne, or of General Patton coming to their rescue with his Third Army. He proudly states that the 4th Infantry Division also played a key role in that most famous of all battles of World War II.[178]

December 6–15, 1944: The order to defend the border between Germany and Luxembourg surprised the men of the 22nd Infantry Regiment. They had just suffered the heaviest casualties of the war and had expected to be withdrawn for a rest. The men were nonetheless glad to be out of the Hürtgen Forest and took each day as it came.

Chaplain Boice: And yet, not a soldier present, from the high command to the lowest private, so much as dreamed that within a matter of days this same group of men would be forced to face all of the power and might of a great von Rundstedt offensive whose plan had largely been based upon the German knowledge that the Fourth Infantry Division, as a fighting organization, should, by all rights, have ceased to exist.[179]

December 7, 1944, Capt. Henley: Raining, but we are warm and comfortable in our CP. Planning complete reorganization of Bn as losses in last fight were 100 percent in combat troops.

December 8–9, 1944: The 4th Infantry Division was ordered to establish a defensive line facing the Sauer and Moselle Rivers along a twenty-five-mile stretch of the Luxembourg-Germany border. Each rifle company was responsible for a mile or more along the lightly defended front. (After Action Report)

Capt. Faulkner: When E Company pulled into Manternach, Luxembourg, from the Hürtgen Forest, we were dirty, muddy, pooped, bashed, and dead tired. It was December 8, 1944. The next morning, two or three of my men arrived at our CP and shouted, "come with us for a bath." So, down the street we went to a bombed-out house and up to the second floor. It was all shelled out, but visible to us all was a tub and a small wood stove above it...making hot water. I took off my boots and jumped into the tub, muddy clothes, and all. As my men applauded, I disrobed and had the first bath since leaving the USA on September 15, 1944. Wow![180]

December 10–15, 1944: Although the 22nd Infantry Regiment was actively conducting patrols, Lieutenant Wilson recalled that a nice house in Luxembourg had been acquired and men were being encouraged to take a few days off. He was surprised when after only one day he was ordered back to his company because of unusual activity at the front.

December 6–31, 1944: Defense of Luxembourg

December 16–26, 1944, Battle of the Bulge: Defense of Luxembourg: German forces attacked Berdorf, Echternach, Osweiler, and Dickweiler simultaneously as German commander von Runstedt launched his large-scale counteroffensive.

Capt. Henley: (12/16) Gostengin, Luxembourg still on outpost duty on Moselle River. Krauts on one side—Yanks on the other. Krauts in northern sector (12th Inf. area) started attack across river this morning on 40-mile front. The 12th Inf. is catching the devil. We have been alerted and may have to help them. Col.

Lanham had a small party at Mondorf for officers and invited a few girls from Luxembourg. Everybody had a good time until we got orders to report back to our outpost as the jerry is about to overrun the 12th Infantry. Dam jerry won't let us have any fun at all. Let them come—we'll open the kraut hunting season again.

December 17, 1944: The full weight of the 212 Volksgrenadier [German infantry] Division was thrown against the 12th Infantry Regiment. The dogged determination of the Combat Team and all the supporting 22nd Infantry Regiment units was the greatest contributing factor in saving the city of Luxembourg and its important installations from being overrun by the enemy. During the day, the Germans extended their advance on Berdorf to the southwest as far as Mullerthal, and from Echternach southwest as far as Scheidgen, but the attack on the Osweiler-Dickweiler area was checked and heavy losses inflicted by the 12th Infantry Regiment. (After Action Report)

Lieutenant Wilson radioed headquarters that Osweiler had been secured by the Second Battalion, rescuing what was left of the 12th Infantry Regiment so they could return to Berbourg. He was told by the battalion's commanding officer they were on their own because the rest of the battalion was under attack.[181]

Capt. Faulkner: ...My Company [E] strength was close to 100 with three lieutenants, one old timer, Lt. Mason, and two new replacements. [Normal company strength is 250 with five lieutenants.] We wondered what the Krauts were up to. No one seemed to know yet for sure. The weather was clear and cold with frost.

Suddenly, burp gun, MG and M1 [machine gun and rifle] fire was heard to our rear. Both Companies G and E reversed direction and worked back through the woods to clean out and contact our people. Situation very tense. I wish Lt. Lloyd was with us...We contacted people alright, but not ours. We got about 20 to 30 rounds of mortar fire which chopped up part of the first platoon, wounding the only aid man and almost got me.

...We had arrived at the ambush point and picked up several wounded from the Hdq. Group and a few dead, among whom was 1st Sgt. Willard. A wonderful man, and I grieved in my heart for him. He had been a D-Day man, wounded twice, but always said, "It's not if you'll get it, but when and how bad." He was right.[182]

North of Luxembourg, in the middle of what would become the German "bulge" into the Allied front line, Private Cassar found himself heading back to Belgium, to Saint Vith.

Pvt. Cassar: (12/17–18) The confusion was everywhere. We are on the move. We left Ubach, Germany and drove 78 miles back through a stretch of Holland and into Belgium...It was worse today than yesterday. There were convoys going in all directions. Seemed as if no one knew where they were going. I knew a while back that Germans were pulling all sorts of crap—changing road signs—posing as American military police sending us to the least strategic places. Gotta give the bastards credit. The road congestion here makes Broadway's bumper to bumper traffic a wishful memory. As we passed, some G.I.'s in the 106th Infantry Division going the opposite way yelled at us. "You guys are heading into Hell. We're on our way out." We were heading into the Ardennes Forest.

December 18, 1944, Capt. Faulkner: At first light we dropped about 12 rounds of 60 mm mortar we couldn't carry anymore. Put it where we thought Jerry was. One came out toward us...Perhaps he was the one that killed my 1st Sgt. The men and I wanted to shoot him—but didn't.[183]

Colonel Kenan sent Lieutenant Wilson and F Company with a platoon of tanks to help Captain Faulkner and his men. Wilson recalled that although they had the Germans surrounded, the Germans had them surrounded more.

Capt. Faulkner: I really prayed and asked the Lord to get us the two miles across those turnip fields and open valley and into Osweiller without fire. He did. It might have been at the expense of several of our tanks who were moving southwest out of the valley, as we could see the gray splotches of Jerry's artillery shells exploding around them as they passed out of sight over the hill. We went on through the mud and into town. Picked our houses in town none too quick, as some big stuff—real big—smashed in the center of town...[184]

December 19, 1944: The weather began to seriously deteriorate with cold and dense fog limiting visibility. Men were woefully underequipped, with little clothing that could protect them from the elements, especially while sleeping in open foxholes. They had no alternative; it was vital that the 22nd Infantry Regiment maintain their positions and push the German attack away from Luxembourg.

Capt. Henley: Ordered to stay in place until further orders. Jerry counterattack is going to town. They have recaptured Krinkelt—St. Vith and now on the Bastogne Road.

December 20, 1944: The 4th Infantry Division was reassigned from the First Army and VIII Corps to the Third Army and XII Corps. Due to the seriousness of the tactical situation and the understrength of the combat teams, sixty-one men from divisional headquarters were sent into combat. In the late afternoon, the Second Battalion of the 22nd Infantry Regiment with two tanks from the 70th Tank Battalion made an attack west of Osweiler toward the high ground west of Rodenhof against an enemy force estimated to be about 400 strong. (After Action Report)

Capt. Faulkner: Wednesday, the 20th saw our "Able Peter" patrol off before first light and we used the rest of the day to perfect defenses…All laid out perfect by 1500 and all changed at 1530 when ordered to move out to northwest at once to woods to attack east to join F Company who were now moving into the woods opposite Rodenhof…[185]

Lt. Wilson: At 4 pm companies E and G jumped off abreast. As they advanced, we clearly heard the staccato barking and ripping of German machine guns; E and G must have been meeting very stiff resistance.[186]

Capt. Faulkner: We saw jerry as he saw us. He was dug in along a belt of bushes above us — 150 yards away. We raced for a wooded draw to his right, got the lead platoon in and opened fire before he did but his MG caught our 4th Pl. [platoon] and heavy MG section attacked and cut them up badly. Just dusk now, and we had a *real* fire fight.[187]

Lt. Wilson: The late afternoon light was quickly fading, and we'd managed to get our foxholes down only about a foot when the men on the outpost rushed in yelling, "Germans!" The Krauts came in right on their heels, firing rapidly. For the next half hour we had some of the toughest small arms fighting I'd ever been in, and I was proud our new men held on so well. These were very stubborn, determined Germans, and they kept right on coming. By then it was completely dark; we couldn't see the enemy and they couldn't see us. We kept firing at the flashes of their rifles and burp guns; every now and then we could see a shadow moving... The firing ended abruptly, and the Germans disappeared.[188]

Capt. Faulkner: The town was getting its worst shelling, very heavy artillery, and screaming mimies [rockets] and burning like a bonfire... Found F Company in the pitch-black dark dug in deep draws with Jerry only a few yards away. We got into some kind of position after midnight. Tried to dig in the stony sides of the slopes. Quit at 4 am and lay in the open holes to sleep. Exhausted![189]

December 21, 1944: At 8:00 A.M., the 22nd Infantry Regiment's Second Battalion continued its attack west of Osweiler toward Dickweiler against very heavy resistance. Numerous counterattacks with enemy infantry supported by small groups of tanks were hurled against the battalion throughout the afternoon, but no ground was yielded. (After Action Report)

Lt. Wilson: The weather suddenly became a vicious enemy... We had no shelter but had to keep pushing the stuff [the snow] away to keep it from melting on us and above all, on our weapons. Then the weather turned bitterly cold at night, dropping to as low as ten below zero... No earmuffs, no hood, no face covering, no scarves. Our hands also suffered with only wool finger gloves…Trench foot became near epidemic…Worst of all were the long, cold nights…No blankets reached us, no shelter halves, no sleeping bags…It must be experienced to be understood.[190]

Captain Faulkner's diary noted "all hell broke loose again" as they moved out the next morning to attack. One of his lieutenants was shot in both legs and the leader of his Third Platoon was killed. That night, a grave digger in the outfit dug an immense hole and covered it with 8-foot logs for protection. They lost more men to trench foot and infections than to injuries. Faulkner was the only officer left in E Company.[191]

December 22–23, 1944: The German front lines extended roughly from Waldbillig, Mullerthal, Osweiler, and Dickweiler, then southeast to the Sauer and Moselle Rivers' confluence. The Second Battalion faced the enemy in positions around the town of Osweiler. Heavy artillery and mortar fire hit their positions and Osweiler. Captain Faulkner noted they were holding hard against the enemy, who was doing the same. Captain Henley wrote that the First Battalion was in reserve in Gostengin. (After Action Report)

On December 23, Private Cassar retreated with the 7th Armored Division from Saint Vith to Manhay, Belgium. Here the division established a defensive line that it held through the end of the year. On Christmas Day he wrote an unsettling entry in his diary:

Pvt. Cassar: I passed many GI's frozen solid in the snow. Noticeably, there was one with his hand raised with a gold wedding band on the fourth finger of his left hand. When I drove by him an hour later, the ring and the finger it was on were gone. What bothered me most about it was that we were behind our own lines. It was no German or Limey that snapped off that finger.

December 24, 1944: On Christmas Eve, the Second Battalion was relieved. On Christmas Day, its men were trucked from Osweiler to Berbourg.

Lt. Wilson: Finally, on Christmas Eve, we were replaced, and my frozen, exhausted men happily marched the quarter mile back to Osweiler. To us it was the best possible Christmas present. There we packed the entire company—all twenty-three of us—on one two-and-a-half-ton truck and my jeep, and we moved back some two and a half miles to the little village of Berbourg that we'd come through earlier.[192]

December 25, 1944, Capt. Faulkner: Relieved by Company K...Sneaked across the frozen fields back to Osweiler, hit the sack for two hours, and started a foot march at 0415 back to Herborn and by truck from there to Berbourg. Approximately 50 men and 2 officers are left in Company E. Rest, it's wonderful![193]

I had never thought about what it would have been like to be continually under fire in deep snow and freezing weather while handling weapons and trying to avoid being killed. One of my father's last comments in his story reminded me that these impossible conditions were not being experienced by everyone.

```
Lt. Sisson: Living in a hotel with French
waiters in the dining room isn't doing
my soul any good as far as soldiering
goes. I even take a bath every other day
```

```
now. It was a real effort at first. Hardly
seemed decent to keep so clean.
```

December 26–31, 1944: The 4th Infantry Division had successfully established a firm defensive line on the southern edge of the seventy-five-mile-wide German breakthrough. General Patton's tanks were able to move behind this secure position on their drive north to relieve the surrounded 101st Airborne Division and other American units defending the vital crossroads at Bastogne, Belgium.

Chaplain Boice: During these vital days, men awoke to fight out of their fox holes; command posts were overrun. Mail clerks and cooks joined the battle and fought valiantly for the life of the regiment. Incredibly, it held!…even though it was over run or surrounded. It fought as it had never fought before, and it survived.[194]

By December 31, 1944, the 4th Infantry Division had suffered 229 officers and 2,939 enlisted men killed in action since landing at Utah Beach. More than 17,000 additional soldiers had been wounded in the 209 days of consecutive combat since the landing.[195]

That same day, according to Boice, the Chicago *Sunday Tribune* reported:

> High praise for the 4th Infantry Division for saving Luxembourg was expressed by Lt. Gen. Patton in a letter today to the division's commander, Major General Raymond Barton of Ada, Okla.
>
> "Your fight in the Hurtgen Forest was an epic of stark infantry combat" said Patton, "but in my opinion, your most recent fight — from December 16 to December 26 — when, with a depleted

and tired division, you halted the left shoulder of the German thrust into the American lines and saved the City of Luxembourg, is the most outstanding accomplishment of yourself and your division."[196]

That was the generals' war. Swede Henley summed up the soldiers' war...

Capt. Henley: (12/31) It stinks as we are still hunting krauts. Goforth went back for a physical checkup. Krauts bombed us today—missed, the sorry bastards.

THIS IS ALL FOR 1944.

35. New Year, Same War

JANUARY 1 – MARCH 12, 1945

January 1–18, 1945: During the first weeks of January 1945, the 4th Infantry Division continued to defend the Osweiler-Dickweiler sector near Echternach. The 12th, 22nd, and the 8th Infantry Regiments were positioned so each infantry regiment defended its sector with two battalions abreast and one in reserve. (After Action Report)

German patrols had draped their men in white sheets to blend into the snow, and the people of Luxembourg provided the American troops with the same. The fighting was not as intense although Lieutenant Wilson continued to note how upsetting it was to train and get to know new recruits, only to lose them as casualties.[197]

Maj. Henley: (1/1) Still in Herborn, Luxembourg sweating out krauts. Received orders for river crossing into Siegfried line—horrible—I fainted.

(1/3) Berbourg, in good ole Luxembourg. Sweating out the jerry and snow...Snow started—cold as the devil.

(1/11) Relieved by 3rd Bn along Sauer River east of Echternach. Still cold as hell.

(1/12–14) Still patrolling across Sauer River and trying to

keep warm. Went back to Luxembourg and took a bath on Jan. 14th. Whiskey ration came in—drinking—awful.

Chaplain Boice: On January 13, plans were formulated, and orders were issued for the Second Battalion to clean out the tongue of land formed by a bend in the Sauer River some 2,000 yards below Echternach. A deep wire entanglement, covered by automatic fire, protected the base of this peninsula...Combat Team 22 was relieved by Combat Team 347 on the 16th of January...[198]

On January 16, Sergeant Cullen was north of the 22nd in the heart of the Battle of the Bulge in Belgium. Fighting from Règné toward Sterpigny and Cherain with the 83rd Infantry Division, he described the conditions faced by the exhausted First Army.

Sgt. Cullen: The weather turned bad, clouds first then more snow and wind. Then the Company was ordered off the road and up into the hills. We were to take and hold the high ground while the tanks maintained the attack along the road. The deep snow was lousy for walking; we were exhausted when we arrived at the top of the tree-filled hill...The snow seemed to cling to the tree branch until someone was directly under it, and then a big lump of wet snow would fall, hit the helmet, and pour down the neck, resulting in a lot of swearing. We were always alert for the sound of a mortar or incoming shells or the enemy himself—but in deep woods, little was heard.[199]

[A few days later near Ottré:] I don't know which regiment of the 83rd Division went through our area, but they passed close to us. We were "armored infantry," and they were "infantry"...We were the troops who were out in front of the rest of the Army.

January 1–March 12, 1945: Back into Germany

I stood by our half-track and watched one 83rd Division Platoon as they sat in the snow waiting for orders to move on. They were a pretty sorry lot… Every GI in that unit had some beard; and all of them had dirty, dark faces with runny noses and weary eyes… Then I looked at my men and I realized that there was no difference in the two groups of soldiers...Later some German prisoners were marched by, and they looked no different, aside from the color of their uniforms.[200]

After a narrow escape from Sterpigny, Cullen was wounded a second time while attacking the village of Cherain on January 18. The shrapnel that almost took his leg was also absorbed by his good friend Roy Plummer's canteen, saving his buddy's life.

> "Very lucky," he said...The Medic came; Roy Plummer and the men patted me on the helmet and said goodbye. They went off to continue the attack and the medics took us down the hill...The shrapnel, I was told, went right through my leg in between the two bones, severing the nerve to the foot and leaving it numb forever. "Very lucky," they said, and I agree.[201]

Sergeant Cullen would not return to combat.

January 19–27, 1945: The Second Battalion remained in reserve at Junglinster until the 4th Infantry Division received orders on January 27 to move north. The First and Third Battalions were fighting along the banks of the swollen Sauer and Our Rivers in the frigid cold and snow to retake territory.

Back on December 20, after the 4th Infantry Division was cut off from the rest of the First Army by the German attack, they had been transferred to the Third Army under General Patton who had established his headquarters in Luxembourg. With a break in the fighting, Lieutenant Wilson took the time to note his influence: K-rations seemed improved, and everyone was required to salute — officers now wore their insignia of rank at all times. Wilson did not mention what Sergeant Cullen did, German snipers targeted officers, a gold bar was easy to spot.[202]

German troops surrender to the 4th Infantry Division troops, January 1945, near the Sauer River, Luxembourg

22nd Infantry Regiment, Prüm, Germany, March 1, 1945

January 28, 1945: The 22nd Infantry Regiment Combat Team was assembled and moved by truck to near Huldange, Luxembourg. The move of seventy miles was completed successfully despite ice, snow, and extremely heavy traffic.

Maj. Henley: Moved out at 0725 from Medernach...Passed thru Bastogne, a complete wreck when full force of German counterattack was thrown against 101st Airborne Division. Hundreds of American tanks lined countryside.

Lt. Wilson: We, who had done our share of attacking small towns, were nonetheless awed by the total destruction of Bastogne...The desperate Germans had attacked Bastogne viciously with what must have been overwhelming force...It was appalling to me to imagine the fighting that must have gone on there. Many bodies still lay where they'd fallen, partly covered by blankets of snow.[203]

January 29–February 3, 1945, Maj. Henley: (2/1) Moved from Huldange to Oudler. Saw a lot of jerry equipment knocked out by Air Corps. Snow melting and everything muddy and nasty. Planning to go back into Siegfried again at same place as before...Russian news sounds good—hope they break Krauts back. The dirty bastard...

(2/3) Crossed IP [Initial Point] at 0745. Received orders to patrol and occupy Bucket [Buchet] tonight...Patrolled into Siegfried line. Found krauts disorganized. CP located in Bleialf.

Lt. Wilson: Back in September, Saint Vith had been a very charming little farming town...All we could see were the jagged outlines of the walls that had once been buildings...Nothing could have survived that bombing…One could not help reflecting on the battles we had fought in the same area in September 1944. We also wondered how many lives had been lost for what appeared to be no gain after almost five months of hell.[204]

February 4–12, 1945: On February 4 they prepared to attack Brandscheid, Germany once again. Major Henley noticed that the men took the same position on the line as on September 13, 1944. Chaplain Boice described an unnerving incident in which troops saw the pants of a soldier who had stepped on a mine back in September still hanging from a tree near the first pill box of the Siegfried Line.

Despite rugged terrain, snow, and rain, the 22nd Infantry Regiment advanced more than 10,000 yards in nine days and breached the Siegfried Line. Traveling on icy and muddy roads that were mined, they took the towns of Brandscheid, Niedermehlen, and Obermehlen, then advanced to the Prüm River against a strong enemy defense. They stopped at the town of Prüm, the strategic Schnee Eifel "focal point." The Schnee Eifel is a wooded plateau region in Germany across the Our River from Belgium. "The Combat Team took 12,373 prisoners during these operations; and approximately 150 Siegfried pillboxes and bunkers were reduced."[205]

This victory came at a bitter cost for the veterans of the Second Battalion. On February 6, Capt. Arthur Newcomb was reassigned from battalion executive officer back to the command of F Company. Under the Army's system of rank and command, a newly arrived replacement major displaced him, despite Newcomb's experience and excellent record extending back to D-Day.

Lieutenant Wilson recalled that Captain Newcomb was very quiet when he arrived to take over command of F Company. He was the only D-Day officer left in the Second Battalion, and he had long deserved promotion to major. Although Newcomb performed flawlessly, the troubling change in his demeanor persisted.

On February 8, F Company attacked Obermehlen and came under serious shelling.

Lt. Wilson: Suddenly the great good luck, the almost sensational good fortune that had blessed me for eight months, abruptly left me there on the open road to Obermehlen. I clearly heard the whistle of the shell and could tell by its sounds that it was falling on me; I threw myself flat on the frozen ground so hard that my chin strap broke, and my helmet flew off...In an instant I heard the shrapnel whipping past and also got a sledgehammer blow on my left foot. I shuddered with the impact as I lay there stunned by the concussion...I wondered if my toes were gone...

I was interestingly watching the medic cut off my boot and expose the wound when I heard Lieutenant Lee Lloyd announce over Colonel Kenan's radio that Lieutenant Wilson had been wounded and Captain Newcomb had been killed. I lay there in tears...Arthur O. Newcomb was my best friend in the Army, and I'll never forget February 8, 1945...

Lieutenant Colonel Kenan, who was also terribly unsettled by the tragedy, later observed that Captain Newcomb was of that rare breed that can act and can inspire others to act courageously on a battlefield...

He was, I believe, an ROTC officer, and he used to talk a lot about the good times he'd had at the University of Wyoming and in the West. He also mentioned quite a bit about his family and their sheep ranch in the hills.

Sometime later, friends at F Company told me that shelling forced them to take cover in buildings as soon as they had reached Obermehlen. Captain Newcomb had been entering a building when a shell hit the doorway. A battle-wise veteran, Captain Newcomb should have taken any cover available, yet witnesses said he had made no effort to protect himself.[206]

On February 10, Major Henley assumed command of the Second Battalion, succeeding Lieutenant Colonel Kenan. I could not help but wonder, although there were no specifics provided by the men, if Arthur Newcomb's death was one burden too many for the lieutenant colonel. It would have been for me.

Captain Newcomb had seen more than his share of combat. He was well liked and generous. My father mentioned him several times. They buddied around England before the invasion, and Newcomb had shared his whiskey ration with Dad when he welcomed my father back to the Second Battalion in Belgium. As far as I was concerned, Arthur Newcomb was killed by a system of rank and promotion combined with combat which, in Sergeant Cullen's words, "Expended men like ammunition."

Obermehlen was taken the same day.

February 13–26, 1945, Boice: ... the Combat Team maintained defensive positions west of the Prum River. For the most part, combat activity was limited to consolidation of positions which overlooked the town of Prum and provided direct observation on the streets of the city. Consequently, movement in the town was restricted, and re-supply had to be carried on during the hours of darkness.[207]

February 27–March 3, 1945: On February 27, the Combat Team received the order to attack across the Prüm River. The goal was to force a bridgehead across the river and take the town of Dausfeld and the two hills surrounding it. After crossing several foot bridges, the First and Second Battalions fought to clear the ridge line. The enemy forces here were determined, sometimes fanatical elements of the German 5th Paratroops Division, as the men referred to them. This Division fought to the death, even when clearly outnumbered and overwhelmed. They showed no mercy and expected none.

Maj. Henley: (2/28) Jumped off at 0530 and crossed river. Casualties heavy from booby traps and mines and machine gun fire. Still fighting 5th Paratroops Div. They are mean bastards. Seized objectives. Attacked—lost 4 tanks fast by AT gun.

The roads were deteriorating with the spring thaw, and railroad beds were being utilized for vehicular traffic. The rapid advance of the enemy during the December breakthrough had left a scene of carnage. Dead soldiers from both sides and animals lay among the twisted wreckage of military vehicles and armaments.

On March 2, Col. Charles T. Lanham was promoted to assistant division commander of the 104th Infantry Division. Lt. Col. John F. Ruggles, a D-Day veteran, assumed command of the 22nd Infantry Regiment. He would rise to the rank of major general before retiring and, along with Chaplain Bill Boice, on the ship returning to the U.S. in July 1945, they became the founders of the 22nd Infantry Regimental Society—WWII. (The 22nd Infantry Regiment Society is a thriving organization today, open to all who ever served with the 22nd Infantry Regiment. Find them at www.22ndinfantry.org).

By March 3, the 4th Infantry Division had established a

bridgehead across the Prüm River for Combat Command B of the 11th Armored Division. The towns of Prüm, Rommersheim, and Weinsheim were cleared by the armored division during their advance through the 4th Infantry Division's forward positions.

March 4–11, 1945: The 22nd Infantry Regiment advanced toward the village of Duppach on March 4, where they battled the German 5th Paratroop Division on the high ground east of the village. James Martin Davis, a 4th Infantry Division Vietnam veteran, told the story of his uncle, Jim Laferla, who fought at Queen's Hill outside Duppach on March 5.

> The 2nd Battalion's mission was to take Queen's Hill with H Company in the lead, despite the fact that it was a heavy weapons company. Sgt. Jim Laferla commanded two squads in the 2nd Platoon of H Company. He was in the point element, moving up the hill on line with the rifle companies. Halfway up the hill, the 2nd Battalion ran into troops from the 5th German Parachute Division. The fanatical German paratroopers, along with elements from the equally rabid Volksturm [national militia], composed mostly of Hitler Youth members, overran one of the rifle platoons, and they began killing the American wounded.
>
> Jim began laying down covering fire with his M-1 rifle in the direction of the German guns while his assistant gunner took over the machine gun and tried to set it up. Jim then attempted to direct the placement of his other gun. Jim managed to get to the gun under a hail of fire. He began firing, and while swinging the gun around, was hit in the abdomen by German machine gun fire…Medic Mickey Lieberman got Jim down the hill where he was quickly removed by stretcher to a field hospital. Lieberman and Lt. Andy Brennan, both

wounded, made it back to an aid station accompanied by a German prisoner.[208]

The capture of the hill allowed the 22nd Infantry Regiment to advance into Hillesheim and on to Niederbettingen, where they met little resistance. German soldiers were withdrawing so rapidly that Americans found foxholes that had been dug to defend the town, but were never used. The secured town offered shelter, and there was no further engagement with the German army as they consolidated their position.

Maj. Henley: (3/6–3/11) Hillesheim set up in a wonderful CP. Reorganizing—washing—clean clothes—paradise. Received order to prepare to move to south. God only knows what is in store for us.

On March 7, the Ninth Army was about 55 miles northeast of the 22nd Infantry Regiment when they found the Ludendorff Bridge still standing across the Rhine at Remagen. "The bridge was seized, the Rhine River was crossed, and for the first time American soldiers poured into the heartland of Germany. The Third Reich's days were numbered." The same day, Sgt. Jim Laferla died a long way from Omaha, Nebraska in Hillesheim, Germany. He was, most likely, the last soldier to be killed in action in the 22nd Infantry Regiment's H Company.[209]

March 12, 1945: The 22nd Infantry Regiment remained at Hillesheim until March 12, when the 4th Infantry Division began assembling to leave the Third Army sector and combat duty. Upon their arrival in France, as part of the 4th Infantry Division, they became a reserve element of the Seventh Army.

Maj. Henley: Moved from Bleialf by 40-8 train to Luneville, France for a rest period. An unbelievable thing.

Lieutenant Sisson's special leave home in January 1945 had allowed him to sit out three ferocious battles his battalion faced. He was a lucky man. Most infantry soldiers were fated to fight until death or serious injury. Wasn't there a way to gauge when a man had done his time? Of all the men fighting on the front line that I followed closely, only Private Cassar, driving supplies for the 7th Armored and several officers assigned to headquarters had not been "hit". I was happy that Sergeant Cullen and Lieutenant Wilson recovered and had assignments at the end of the war that did not put them in harm's way. Too many young men I had met in my father's story had died a long way from home.

Now that I was nearing the end of my journey, it seemed to me that we would all do well to know what combat demands of a human being. When trying to fathom how the soldiers of the 22nd Infantry Regiment continued the fight in the Hürtgen Forest, Robert Rush had summoned the words of a Union soldier during the Civil War:

> The truth is, when bullets are whacking against tree trunks and solid shot are cracking skulls like eggshells, the consuming passion in the heart of the average man is to get out of the way. Between the physical fear of going forward and the moral fear of turning back, there is a predicament of exceptional awkwardness, from which a hidden hole in the ground would be a wonderfully welcome outlet.[210]

On leave a month before he was wounded, George Wilson had this to say about his own colleagues who had not experienced the actual fight:

...I found myself bound for Paris, for some reason the dream city of almost all GI's...I couldn't believe the food, having almost forgotten that such lavish, delectable victuals existed...What did bother me, however, was the kind of life our rear echelon troops, especially the officers, seemed to be leading. In a fantasy I was sending them up for a tour of front-line duty just so they'd know there was a war somewhere.[211]

36. The Final Chapter

MARCH 13–MAY 30, 1945

In September 1945, the WWII yearbook of the *4th Infantry Division's 22nd Infantry Regiment* was published for the men of the regiment. Pictures and narrative told the story of the 22nd Infantry's significant contribution to the war effort. This is how the yearbook described the final months of the war for my father's regiment in Europe,

> The attack continued, and day by day the gains became larger and larger, movements being made sometimes by foot, more often by motor. At various points throughout the advance, the German resistance stiffened, and vicious local fighting ensued...Five, ten, fifteen miles a day, the attack surged and eddied like a storm-swollen river that has broken its banks; death and destruction lay everywhere in the wake of the Allied effort...and from early morning to late at night the attack continued, town after town falling to the 22nd Infantry Regiment, "Surrender or be destroyed—and be quick with your answer!"[212]

In 1994, John C. Ausland, a captain who landed on D-Day with the 4th Infantry Division's 29th Field Artillery Battalion, wrote an article for the *International Herald Tribune*. He offered his own views on the final months of the war:

One reason people born after World War II find it difficult to understand why the final days of the war were so destructive is that they do not realize how angry we Allied soldiers had become—and to some extent we still are.

Once our forces crossed the Rhine, it was clear that Germany was doomed. But Hitler, in his madness, vowed to fight on to the finish.

Anger became outrage and horror as our forces began to overrun concentration camps.[213]

March 13–14, 1945: In mid-March, the men of the 22nd Infantry Regiment had yet to start their advance through Bavaria. They were leaving Bleialf, Germany and being transported to Lunéville, France for rest. The 4th Infantry Division was now a reserve element of the Seventh Army. It was the first time that the 22nd Infantry Regiment had not been in direct contact with the enemy in 199 days.

[Editor's Note/4ID Historian: This marked the first time in 199 days (that's what the 4ID history says in our yearbooks,) but to the best of my understanding, this is the first time since D-Day we have not been on the front lines. I believe we were constantly in contact with the Germans for 279 days until we got this break.]

On March 11, First Lieutenant Sisson returned to the European Theater of Operations and was re-assigned to the 22nd Infantry Regiment. He had yet to join them.

Maj. Henley: (3/13) Moved from Bleialf by 40-8 train to Luneville, France for a rest period. An unbelievable thing. Set up CP in Domptail in Mayor's home. E & H companies in Xiffevilles [Xaffévillers]—F & G in Menarmont.

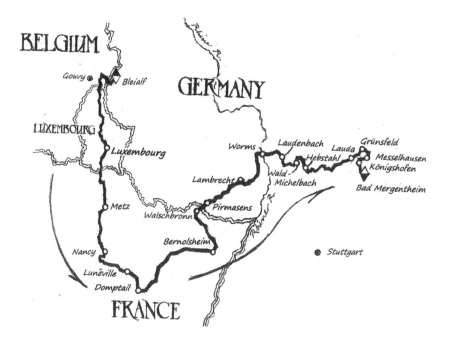

March 13–April 9, 1945: Across the Rhine River

March 14-20, 1945, Lt. Sisson: (3/14) Dear Mother and Dad, I wouldn't say the return to the States has exactly inspired this bunch for more war—but it has made them sure of one thing—to get back there as soon as possible and to stay for keeps...It goes without saying it was wonderful beyond description to be back. I get a laugh now, when I think how little time it took to get back into a settled way of living—and how little time it takes to get out of it again. At least the army teaches you to take it as it

```
comes. And it does. Hope you're well—love
to all—John
```

Maj. Henley: (3/15) In Domptail. Still training—resting—reorganizing—weather has been beautiful.

March 21–24, 1945: Chaplain Boice wrote that training and rest continued. The 4th Infantry Division's next assignment would be to cross the Rhine River and advance rapidly southeast to the Brenner Pass to isolate enemy units in Italy from those in the Bavarian Alps. Although this was not part of the main attack to defeat the German army, it would disrupt communications and transportation and drive the enemy toward their southern border with Switzerland.[214]

March 25–29, 1945: The 4th Infantry Division began to move toward Germany and the Rhine River. Memorandum Number 13 was issued concerning Looting and Fraternization in Occupied Territory. *There was to be none.*

Maj. Henley: (3/25) Moved out at 0830 through Baccarat to Bernolsheim. The weather was beautiful and rest still good. Visited Strasbourg. 4th Division moved to operational control of the XXI Corps.

(3/26) Moved out at 1416 for Lambrecht. Passed thru Siegfried defenses at Pirmasens and it looked nasty. Drive down thru the valley was beautiful and a good road. Air Corps had killed a lot of horses. War news sounds good—everybody is on the go.

(3/27) CP in Lambrecht awaiting movement orders. As per usual it is raining again since we are back in Germany.

Lt. Sisson: (3/29) Still in the Replacement Camps. We got a train ride in box cars the last time—and it was a beauty. About enough room to be perfectly comfortable—if you don't try to take a deep breath or try to straighten out your legs…We were kings on the way home—but we're just pretenders sneaking in the back door on the way back. The looks of jealousy we get from everyone who finds out we've been home have me convinced, though, that we must have had something very enjoyable.

March 30–31, 1945: The Germans continued to fall back in apparent confusion and kept well ahead of the advancing troops, avoiding all but minimal contact. Resistance in the form of scattered rifle shots and an occasional hand grenade thrown by soldiers in civilian clothes or by bona fide civilians in a burst of anti-Allied feeling characterized the period. (After Action Report).

On March 30, the 22nd Infantry Regiment crossed the Rhine and moved to the rear of the Seventh Army's advance. Resistance became more common; they began to face pockets of German soldiers protecting their homeland.

Maj. Henley: (3/30) Crossed Rhine River at 0530 on Pontoon Bridge at Worms (what a relief to cross that bastard). CP set up for night at Michelbach and was it a honey—electric lights, radio, and everything. Kraut frauleins look good but no fraternization. Received orders to move to Hebstahl by foot.

On March 21, 1945, Private Cassar was well north of the 4th Infantry Division near Röttgen, Germany. He was surprised by the relatively friendly reception from the residents. He did not give a location when he made this entry a week later.

Pvt. Cassar: (3/30) Last night our Combat Command captured 200 something prisoners. At the rate we've been taking in prisoners, I doubt there will be any left by the time we meet the Russians. Since that first bunch, there have been thousands of German prisoners; many of them from 12 to 14 years old and others old enough to be their grandfathers. Pretty sure they'd rather be captured by us than the Russians. They are known to be vengeful and brutal. Payback is hell! A couple of days ago my company lost two trucks, one with gas, one with ammo. The ammo truck was hit by a 75 MM high explosive shell and was blown sky high. Both men in the truck were hit in the legs by shrapnel. One of them was picked up and sent to the rear shortly after. The other owes his life to a German medic who worked on wounded for hours and then surrendered to an American MP. Guess all Germans aren't heartless.

Maj. Henley: (3/31) Moved to Hebstahl. Closed at 1600 hours. CP in Beerhall and Inn. Received orders to move out at 0600 in morning for 50-mile attack. Company of tanks attached. Moving for speed and time now.

April 1, 1945, Maj. Henley: April Fool and Easter Sunday (Would like an egg for Easter). Moved out on 50 mile march at 0615. Arrived at Lauda at 1400 and found bridge across Tauber River intact. Pushed on to Gerlachsheim and ran into hell. Captured or killed 350 krauts. Pushed on to Grunsfeld. Captured or killed 125 more krauts.

```
Lt. Sisson: Easter morning and all is in
order. Even had fresh eggs for breakfast.
One meal with fresh eggs is generally
comparable to a pointless roast beef from
Andy…The fellows are around the radio,
sticking pins and moving front line mark-
ers on the maps all day. What the out-
come of all the present activity will be
is hard to say, but easy to dream of…The
Stars and Stripes had a piece saying you
would have to walk a mile to find anyone
in the States who would take money against
the war being over in a couple of weeks.
```

April 2, 1945, Chaplain Boice: In talking to a couple of civilians living in the town [Königshofen], it was learned why the town had been burned. German SS Troops several days before had found civilians possessing the white flags of surrender, which they intended to display upon the arrival of the Yanks. The SS Troops immediately poured gasoline on the houses and set fire to them, telling the civilians that they would be shot if they ever attempted to surrender. Hundreds of people were homeless, and children gathered in the school building, crying that their homes and parents were gone.[215]

April 3, 1945, Maj. Henley: Helped 1st Bn get unhitched from the krauts. Fighting a bunch from NCO [noncommissioned officer] school and are they mean. They fight to the end.

April 4, 1945: Despite the stiff resistance from SS troops well-entrenched in high wooded areas, the Second Battalion of the 22nd Infantry Regiment occupied the town of Messelhasen. (After Action Report). Chaplain Boice wrote, "the enemy were found to be paratroopers and as fanatical as any troop of German soldiers we had fought. They were young, from fifteen to nineteen, but they fought with a fanaticism of which we had read but seldom had met." One injured Hitler youth blew his own head off with a grenade rather than let a medic help him.[216]

Leaving the Königshofen Ridge and advancing toward Bad Mergentheim, Boice recalled that "the German dead lay piled like cord wood over every conceivable defense terrain feature."[217]

April 5–9, 1945: The surrender of Bad Mergentheim to the Second Battalion's medical officer, Captain Herrick, became an oft-repeated story. Even my father wrote home about it. "Doc" Herrick was leading a medical patrol that was seeking to secure the release of wounded American soldiers when the town surrendered to him. "To much the surprise of the patrol, the Burgomaster stated the German civilians in the town of Bad Mergentheim persuaded the German soldiers to surrender or withdraw. The civilians were also responsible for keeping the bridge intact."[218]

Maj. Henley: (4/5) CP in hotel at Bad Mergentheim. What a place—second to none. Captured big radio station (propaganda)...living like kings and enjoying life...a very pretty town—28 hospitals—the krauts aren't hurting for anything. Hospital had cure for crud, but it did not work on Ruggles or Col. Lanham...too far gone.

April 10–May 30, 1945: Advance into Bavaria

```
Lt. Sisson: (4/6) You can write to the 2nd
Bn and be safe now. I'm still with the Bn.
Hq. and still S-2. We have a new battal-
ion commander—Maj. Henley. Helen Bridgeman
probably knew him as Swede Henley, then a
Captain [Third Battalion]. Found a lot of
my old friends here…Your cookies -inci-
dentally—were greatly enjoyed by the men
during my absence and they thank you.
```

April 10-18, 1945: Major Henley recorded taking 21 towns from Markelsheim to Lohrbach, with a memorable command post at a castle in Niederstetten where the prince and princess still occupied the premises. He noted that if the mayors acted "all right," their towns were saved. Surrender, however, was still treason, and Adolzhausen, like Königshofen, was set ablaze by withdrawing SS troops.

Lt. Col. Ruggles: It is with the highest praise that I commend the officers and men who so gallantly fought the battles at Lauda, Konigshofen, and Bad Mergentheim. Words of praise could never express to each individual the sincere appreciation I hold within my heart for this great victory.[219]

Pfc. Tarkenton: Along about April 1945, we were going through southern Germany like cow chips through a tin horn...the command element of the Third Reich was breaking down...Men were still being killed and wounded. I was with many of my brothers when it was their time to leave us...Then I would try to blink away the tears, all the while trying to breathe again with the severe shortness of breath, the excruciating pain of loss of a trusted friend and the tightness in my chest.[220]

April 19–23, 1945: More than 20 towns, from Lohrbach to Dewangen, were captured during this period according to Henley's diary. He wrote that the 22nd Infantry could still face determined resistance at times, including nebelwerfer [weapons capable of delivering large quantities of gas or high explosives], heavy mortar, artillery, and heavy small arms fire. They advanced 50 miles during this period.

Lt. Gordon Gullikson, Company A, First Battalion, 22nd Infantry Regiment: We kept pushing the Germans in retreat and came to a small river. Goforth called Regiment and told them he would have to find a way to get across. I said, "Colonel, there's a bridge about a half mile downstream." He said, "Gordon, the war is about over. If we close on the Germans, they will have to stand and fight. We will just keep pushing them back as they retreat. No one wants to be the last man to die for his country."[221]

"Crailsheim, April 20, 1945"

April 20, 1945: In my father's scrapbook there were black-and-white photographs from this period of his colleagues in jeeps or standing together in comradery. A lone picture of what was essentially a pile of rubble was labeled "Crailsheim." It seemed out of place and confused me until I read:

Chaplain Boice: One of the things which Americans will never forget was the order requiring every German house and building to fly a white flag or be destroyed. An advance patrol was usually sent into the town ahead of the combat troops and artillery, and the Burgermeister was informed that if the Germans defended the town or if American soldiers were harmed within its precincts, it would be burned to the ground.

If, on the other hand, the civilians did not interfere and forced the withdrawal of German troops to the other side of the town, it would be spared. In almost all cases, the Germans of Bavaria were anxious to have their homes saved, and the white flags went up at once. Usually, a time limit of thirty minutes was set.[222]

At Crailsheim. the American 11th Armored Division had withdrawn after a fierce battle ten days earlier. Now the 22nd Infantry Regiment's First Battalion was ordered to take the town of 10,000 inhabitants. Lieutenant Jones and two soldiers were chosen to deliver the terms of surrender to the city officials. As they were leaving, apparently successful, they were shot in the back and killed.

Chaplain Boice: When the battalion commander was informed of what had happened, he issued orders with a tenseness and concealed fury which communicated itself to every man in the battalion. Instead of ordering that the city be attacked, he called for artillery fire, and he personally adjusted the mortars, ordering them to fire white phosphorus and to burn every building in the city. These orders were carried out...If occasionally the smoke and flames blew away and a building was revealed still standing, the Colonel adjusted the mortars and soon the building was in flames...A city died and remembered a Lieutenant named Jones.[223]

When the battalion finally entered the city, the men paired up and, still boiling with rage, prepared to seek out and kill the enemy.

Chaplain Boice: While the fighting was going on in Crailsheim, the Second and Third Battalions moved up to secure the flanks and rear. The men of these battalions could clearly see the burning reflection and they, too, understood what had happened.[224]

April 22–23, 1945, Capt. Ausland: Ellwangen was an SS military training center with a model defense. When the commander of the 29th Infantry Regiment suggested by telephone that the German commander avoid a useless battle, he was rebuffed with an obscenity. On the night of April 22–23, the 4th Infantry Division and supporting corps artillery fired 1,500 shells into the town. We used delayed fuses so the shells would penetrate buildings before exploding...After our bombardment began, the German commander had second thoughts. Changing into civilian clothes, he and his men quietly disappeared.[225]

Years later, Ausland returned to Ellwangen. He learned that, as the bombing continued that night, the residents were unable to find the key to the church where they had hidden their white flags. The residents retreated deep into cellars, praying for survival. The next morning, after the attack ended, they promptly surrendered.

April 24–29, 1945: Major Henley no longer bothered to keep a list of seized towns—both he and my dad referred to their rapid advance as a "rat race." The final surge sped across 180 miles. On April 25, they crossed a bridge over the Danube River near Lauingen, then assembled near Gundremmingen. On April 29, the 22nd Infantry Regiment crossed the Lech River at Graben and secured a defensive line along the Amper River in the vicinity of Schöngeising.

Chaplain Boice: The advance crossed the Kocher River through Zang into Heidenheim. Heidenheim was a beautiful city situated in the low, rolling hills of southern Germany. The Third Battalion was the first element of the Combat Team to move into the city. This they did the evening of April 24th, and by noon the next day the entire 4th Infantry Division had moved into the city.[226]

Maj. Henley: (4/23-30) Rat race continued—gaining up to 20 miles a day. Traveled the autobahn—saw gobs of jet planes along road.

```
Lt. Sisson: (4/25) If I tried to fill in
all the details, I'm afraid it would take
a volume. My main joy is the fact that
the S-2 gets a jeep. If I had to walk
half this far, I would be all washed up.
```

Pfc. Tarkenton: By the last week of April, we were less than a hundred miles from the German-Austria border and closing fast. Heavy numbers of those of the German military were deserting their units to try to return to their homes in peace. Almost every night now, small groups of approximately six to eight men, in ragtag civilian clothes, would unwittingly stumble into our positions...Always they denied they were deserters. They were simply farmers going home. Sure, farmers always go home at three in the morning...We knew they were lying, and they knew that we knew, but who cared?...They were sick and tired of a war they did not want and did not start...They were polite enough gentlemen and grateful to still be alive...They feared us, but they were more fearful of falling into the hands of the Russians...If we had food, we would give them some of ours.

After all, our food came from the last German town we had just secured.[227]

Capt. Ausland: A few days after Ellwangen, I visited a concentration camp that had been taken near Landsberg. It is hard for me, even today, to describe what I saw there without crying. Hundreds of bodies, I wrote to my parents soon afterwards, were laid out in neat, efficient rows. Some were burned. Some were shot. Some had been tortured and maimed. Others may have just died.[228]

Dachau, April 1945

Landsberg was a subcamp of the Dachau Concentration Camp. The 22nd Infantry Regiment was not involved in liberating the concentration camps near Munich. However, a picture of a railroad car filled with grotesquely twisted bodies in the souvenir roster for the regiment makes it clear that the soldiers, like Captain Ausland, witnessed evidence of the atrocities. The picture is labeled, "Dachau Concentration Camp inmates. Much of Bavaria had to be seen to be believed."

April 30, 1945: The 22nd Infantry Regiment crossed the Amper River near Fürstenfeldbruck just outside Munich on the morning of April 30 and continued forward against weak resistance until they were halted along the Isar River near Schäftlarn by destroyed bridges. They had passed within 12 miles of Dachau. The concentration camp had been liberated a day earlier.

Pvt. Richard Marowitz, 42nd Infantry Division: Now, it's not uncommon to smell death...you smell different things depending on what's going on; usually, it was farm animals that were killed...It just smells like hell. As we got closer to Dachau, we didn't discuss it. We were used to that. We all thought the same thing. We never talked about it, but later on afterwards, after the fact, we realized that we all thought the same thing...we're coming to another bombed out farm with a bunch of dead animals...The prisoners were just walking skeletons, and they just dropped where they stood and died. There were piles of bodies, of bodies that had been gassed and readied for the ovens. Some of them still lived because those boxcars were brought to Dachau to burn those bodies...And the smell was not a farm, it was Dachau we smelled miles before we got there.

Well, as soon as I saw what the camp was...you really can't describe it. You really can't. It's not possible.[229]

May 1, 1945, Maj. Henley: Jumped off on another rat race and captured Gmund on lake at foothills of Alps. Captured 3000 prisoners. Want to go home—but not to C.B.I. [military designation for the China and Southeast Asian or Burma-India theaters].

May 2–3, 1945, Chaplain Boice: The resistance just north of

Gmund was the heaviest reported and consisted of small arms, mortars, and artillery fire. By late afternoon, the Combat Team's advance to the southeast covered over twenty miles…The Combat Team stopped for the night…the Second Battalion in Gusteig, Duenbach, Festenbach…little did they realize that this day had been the last day of actual combat for them in the European Theater of Operations…

On May 3, the Second Battalion swept through their zones but encountered no enemy. They moved near Holzkirchen. "Totals compiled by the S2 Section showed that the 22nd Infantry Regiment had in the past two days processed 3,270 prisoners."[230]

May 4, 1945, Maj. Henley: (5/2–5/6): Relieved by 101st AB and moved to vicinity of Nurnberg [Nuremberg]. CP located at Roth. Passed thru Munich and you can mark that one off the books. CP located in Baron's home. Guarding bridges and restoring law and order. Was the Baron mad when we ran him out of his house, the poor bastard.

May 7, 1945, Maj. Henley: Sitting on my can and the war has ended. V-E DAY MAY 7, 1945, AT ROTH, GERMANY 11 months 2 days of combat for the 22nd Inf.

> Lt. Sisson: At long last, or something to that effect, it looks like things are getting into a fairly settled sort of condition. At last call on the radio, the whole thing is over. Alles kaput, as the Germans say…I hit the outfit about the time they took off on a large rat race after the remnant of the Kraut army…As

> you have been able to tell from the reports that have come through in the paper, the Fourth is or was, engaged in cleaning up things in southern Germany. Perhaps you saw the report about the medic liberating Bad Mergentheim—that was our battalion surgeon. All that has been over for some time now. As far as the future is concerned, I think you can rest at ease.

May 8, 1945, V-E Day, Maj. Henley: They announced it. Ruggles got over the crud. Tommy Harrison stopped stuttering. Geo. Goforth lost all of ailments and the psycho said he was normal. Swede was laughing as he was never hit—poor shots—the dirty bastards.

> Lt. Sisson: Frankly, all the peace news seems more or less of an anti-climax, so if I'm not enthusiastic enough about it, don't let it worry you. In the meantime, don't let this VE business get you down, but you may open a bottle of beer for me.

Chaplain Boice: There was no revelry last night, no drunkenness, no shouting, no flag-waving, no horns blowing; there was a sober realization that it was all over, at least so far as Europe was concerned...We simply thought of the hundreds and hundreds of our friends who had given everything that they had in order that we might see V-E Day.[231]

When the war in Europe ended on May 8, 1945, the 4th

Infantry Division had participated in all the campaigns from Utah Beach through Germany. Five more battle streamers were added to the 4th Infantry Division colors, and its soldiers wore the campaign stars of Normandy, Northern France, Rhineland, Ardennes, and Central Europe.

During eleven months of fighting, the division had suffered almost 22,000 battle casualties and more than 34,000 total casualties, including over 5,000 men who were killed or died from injuries. For 199 straight days, the 4th Infantry Division was in constant contact with the enemy. They had advanced more than 1,600 miles.[232]

> May 12, 1945, Lt. Sisson: So we are sitting in a lovely old 15th century castle which some thoughtful kraut modernized for us, figuring our points, and meditating on whether spring can really be beautiful in a non-fraternal country…For the first time since I got back, though, I can take a bath in hot water whenever I want…The thing that is amazing is the number of ex-prisoners, slave laborers, etc., all called displaced persons now. There are literally hundreds of them on the roads each day in our area, and various camps in town are full of them. They walk, ride, or struggle along, with every sort of contraption imaginable…Men, women, and children in groups from two to three to a hundred or more can be seen every day going somewhere. Sometimes they

```
have made a flag of their country, which
they carry at the head of the column...We
have been lucky here, in that we have no
responsibility for them, outside of per-
mission to go through the town.
```

On May 14, the 22nd Infantry Regiment moved from Roth to Dinkelsbühl, which was and still is a beautiful walled German village. On May 10, my dad had proudly written to his parents that Major Henley made him S-1, Commander of Headquarters. He was displaced by a higher-ranking transfer several days later and became S-2 again. This was solely a matter of administrative function now, no longer life and death.

The war in Europe was over, but I would never forget the men of the 22nd Infantry Regiment and what their generation confronted and finally brought to an end.

Lt. Walker, J. Sisson, Maj. Swede Henley

Chaplain Bill Boice

Part Eight

THE DEATH OF A SOLDIER

"Freedom has a price the Protected will never know…"[233]
—**Combat Infantrymen's Association,**
Memorial Walk of Honor,
National Infantry Museum

Chateau de Fontenay 1944 Battlefield, Normandy, France, John Sisson, 1979

This somber spot brought back vividly that Ozzie Wirtzberger's squad had taken a German anti-tank gun with its crew, just behind the ivy-covered pillar. Ozzie was lying in the grass just in front of the wall. Ralph Carter was wounded, lying across the wall under the trees. Jenkins and I dressed his wound at the chateau and Jenkins stayed by him for over an hour. We couldn't get litter bearers up for either him or Wirtzberger.

   I recall trying to get Lt. Camper to set up a light machine gun at the corner of the wall. The Krauts were dropping mortar and light arms fire into the trees. I was hit while trying to move up along the edge of trees and the open field behind the pillar.

37. "Himmler's" Home

SPRING 2023

```
I was the officer in charge of securing
Heinrich Himmler's home at the end of
the war. There was a large, gold embossed
Mein Kampf on a pedestal next to his
desk. It was too large for my pack, so I
took the smaller volume from his desk.
When I got back to headquarters, I real-
ized it could go home with my commanding
officer or me. I decided to keep it.
```

I had stood at the wall of silence that death erects, and I had tried to peer around it by listening to my father and the men of the 22nd Infantry Regiment he served with during World War II. For two years I had searched for one man's actions at the end of a conflict that involved millions of people—and I believed I had found it within the soldiers' war. My father's familiar family tale and Mr. William's startling statement now mingled with the stories of my dad's comrades.

With all due respect, Mrs. Marshall, I believe your father lied to you.

Among my father's wartime letters home, there was a postcard of a pleasant lakeside village, Gmund am Tegernsee. The text was in German, but on the back my grandmother had written, "Himler's Home." Although it did not show the actual house, it seemed reasonable my father had shared the postcard as well as the story with his family. I decided to honor my grandmother's evident pride in her son by choosing her misspelling to title the chapter where I proved for myself what she would never have questioned—her son's integrity.

My problem was not establishing the location of Himmler's residence. To verify my father's story, I needed to place him there in early May 1945. As the old saying goes, "Getting close only counts in a game of horseshoes." I had embarked on a journey to follow him with the Second Battalion of the 22nd Infantry Regiment from Utah Beach to Bavaria to see where we would end up.

To my happy surprise, near the end of my journey I found a significant piece of evidence in the diary Major Swede Henley had kept throughout the war. On May 1, 1945, he wrote, "jumped off on another rat race and captured Gmund on lake at foothills of Alps. Captured 3,000 prisoners." His entry was significant because he was the commanding officer for the Second Battalion of the 22nd Infantry Regiment.

First Lieutenant Sisson served as the Second Battalion's S-2, Intelligence Officer. He reported to Henley. The job of the combat battalion S-2 was to provide his commander with enough timely and useful information to allow him to make critical decisions on the battlefield. The S-2 had access to and the support of higher intelligence channels, but most of their information came from on-the-ground scout platoons and carefully maintained maps and reports about enemy disposition. My father would have "jumped off" with Henley and the battalion on May 1 for the drive to capture Gmund.

Piecing together bits of information from Chaplain Boice, After Action Reports and Major Henley's diary, I was able to develop a reasonable picture of what occurred next from May 1 to May 4, 1945.

On April 30, the 4th Infantry Division and its 22nd Infantry Regiment arrived south of Munich near Schäftlarn to find a "traffic jam" created by the destruction of the bridges crossing the Isar River. On May 1st, two companies from the 22nd Infantry's Second Battalion crossed the Isar River at Wolfratshausen, where several intact bridges had been discovered. This was the beginning of the military action (or rat race) Henley referred to. The remainder of the regiment, including Major Henley and my dad, crossed the Isar River on the morning of May 2.

Chaplain Boice described that day. The regiment advanced south toward Tegernsee, and "the Second and Third Battalions swept through their zones of action, meeting only scattered token resistance." He noted that the resistance just north of Gmund was the "heaviest reported that day." He wrote, on the night of May 2, "the Second Battalion was deployed at Festenbach, Duenbach and Gusteig. The First Battalion was at Holzkirchen and the Third Battalion in Meisbach."[234]

I found Festenbach was on the map and decided Duenbach was most certainly a mistyped Dürnbach. I could not find Gusteig, though there is an inn in Gmund that includes "Gasteig" in its name. Festenbach and Dürnbach form a rough line from north to south with Gmund am Tegernsee.

May 2 would be the last night of combat for the 22nd Infantry in Europe, although "mop-up" operations continued the next day. Chaplain Boice wrote that by May 3 the Regiment had processed 3,270 prisoners in the past two days and that on morning of May 4 the 22nd Infantry Regiment had been deployed to the Nuremburg area, which agreed with Henley's diary entry.[235]

I went back to the last map Ann had drawn for me. The 22nd Infantry Regiment's Second Battalion movements during May 1–3 were confined to a relatively small area. It was 28 miles from Schäftlarn to Festenbach, Germany. Festenbach and Dürnbach were on the outskirts of Gmund am Tegernsee. This meant that from May 2 to May 3, my dad was no more than an hour's walk, or a 10-minute drive, from Sankt Quirin, Gmund am Tegernsee, the site of Himmler's home.

I had discovered a letter written on May 30, 1945, confirming the specific location of Himmler's home and the army's interest in it. Special Agent Lawless had listed materials secured from the homes of Heinrich Himmler and Nazi Party Press Chief Max Amann in St. [Sankt] Quirin, Gmund am Tegernsee. The artifacts were received from Lawless on May 25 and placed at the document center for the Seventh Army in Munich. The letter does not say when they were confiscated, and there is no *Mein Kampf* among the books that were listed. The letter indicates the seized documents were important to understanding Himmler's motives.[236]

It would appear from the Lawless letter that by May 25, 1945, both Volumes I and II of his personal *Mein Kampf* had been removed from Himmler's study (if both had indeed been there) as had the large, gold embossed presentation volume my father had mentioned. In his memoir, General Bradley recounted an amusing and pertinent anecdote:

> Among the troops in Normandy, Cherbourg's strategic worth was soon overshadowed by the wealth of its booty, and it was there that the term "liberate" came into popular use in the army. Von Schlieb-

en's forces had thoughtfully stocked their underground shelters for a prolonged campaign—and for what could have been a historic binge. And while they scrupulously complied with orders to demolish the port's installations, their [German] soldiers' hearts rebelled at the sacrifice of destroying or spilling good wine and brandy. As a result, we fell heir not only to a transatlantic port but to a massive underground wine cellar as well.[237]

My dad wasn't there for the fall of Cherbourg, but he would take to "liberating" like a duck to water for the rest of his life. Securing the *Mein Kampf* was not the only time he participated in this Army tradition; he also had a German dagger and Luger. He had even written to General Ruggles,

```
"the copy of Mein Kampf which I liberat-
ed at Heinrich Himmler's summer home…The
volume was signed Heinrich Himmler—1925."
```

There was no longer any reason not to believe my dad. I was convinced that between May 2 and 3, 1945 my father led a scout platoon that secured Nazi leader Heinrich Himmler's residence in Sankt Quirin, Gmund am Tegernsee. They may have been ensuring Himmler was not hiding in the house. This was part of the mop-up mission assigned to his battalion in that area.

The large, gold embossed *Mein Kampf* in Himmler's study caught my dad's eye. He found a smaller version of the same book on the desk. The notations and underlined sections might have interested him as the S-2 gathering information. But he also recognized loot when he saw it—Heinrich Himmler had signed his name to the volume. He picked it up and put it in his pack along with any other information that he gathered. He had no idea it was Volume I and that there was a Volume II.

The S-2 is required to report all information to their commander, in this case Major Henley. My father's family story even suggested that his commanding officer was aware of his actions. It is possible that some of the documents that made their way to Special Agent Lawless were in my father's report. But Lieutenant Sisson had no problem with bending and breaking rules, as I learned in his letters and war narrative. Liberating war booty was acceptable conduct. If the book entered the Army system, it did not stay there—it went home with him. And his commanding officer? From what I had learned of the engaging Swede Henley, he seemed unlikely to have objected. After all, contrary to regulations, he kept a diary from D-Day to the end of the war.

Did my father recognize the book's symbolic value? Its words had inspired its owner, one of the masterminds of the catastrophic events he had personally experienced. Was he aware of the horrors of the death camps when he liberated the book? My father took that answer to the grave. But I believe he understood. He kept the grim paperback entitled *Himmler* with *Mein Kampf, Volume I* while it was in his possession.

The conclusion of my father's story caused me to revisit my conversations with Charles Williams. He had been certain his father said he had both Volumes I and II at the end of the war. This could not have been the case from what I had learned, and it made me wonder if Charles had misunderstood his father. After all, I still had an unanswered question or two. I was all too aware that in 1991, just as my dad was preparing to resolve the "matter of the *Mein Kampf*" (as he wrote to General Ruggles), he lost his fight with cancer. I will never know the full story or the outcome my father intended:

I did, however, understand one thing. Charles Williams had helped my family find the right home for the book. I wish he had been in Mr. Morgenthau's office the day I presented my mother's gift to the museum. Our fathers had given their families the precious opportunity to take a stand against bigotry and hatred through our respective contributions to the Museum of Jewish Heritage — A Living Memorial to the Holocaust.

Instead of questioning our fathers, perhaps Charles and I should have considered them agents of a higher good. Seemingly accidental, unrelated events had insured that Heinrich Himmler's personal copy of *Mein Kampf* had been prevented from being glorified and sold as Nazi memorabilia. Instead, both volumes were in a museum that testifies to the truth about one of the darkest chapters in the carnage of World War II. A museum where dead innocents will forever be honored and where the living will always be reminded to "never forget" the need to confront and reject the conniving and beckoning voice of evil.

I remembered an article I found written by Tony Hayes, the son of a World War II infantry soldier. Tony recounted how his dad had served in most of the major campaigns in Europe with the Seventh Army's 45th Infantry Division. On April 29, 1945, his father's battalion, and the 42nd Infantry Division liberated the Dachau concentration camp. Tony's dad had a few stories he told, mainly about war's ironies, but Tony was emphatic about one thing, "He never spoke about Dachau, ever."

Tony wrote movingly about a tattered photograph of the skeletal remains of the starved and murdered Dachau inmates that he found in his father's wallet after his death. "Why that one of all the pictures he had in his old scrapbooks?" Tony wondered. Years later, he visited Dachau and met a veteran and his wife at a restaurant. Tony asked the man if he was at Dachau during the war. The old warrior hesitated for a moment, and then with tears glistening in the corners of his eyes he flicked a picture like Tony's dad's on the table. Tony excitedly asked the man,

> "Why? Why have you carried it for so long? To remind you of the horror of Dachau, of what had been done there?"
>
> His face carried the faintest of smiles as he shook his head. "No, son, to remind us of the horrors that we are capable of, to remind us not to go down that road again."

Tony ended his story with these words, "Dachau is still with us, and I hope the legacy left by our fathers always will be."[238]

I finally felt I understood that lone haunting picture of Holocaust victims in the 4th Infantry Division 22nd Infantry Regiment roster. And my father's picture of his regiment's retribution at Crailsheim—a pile of rubble so markedly at odds with the other pictures of smiling comrades in his scrapbook. The soldier's war had helped me realize that individuals who have witnessed or been part of the horrors that we are all capable of, don't need more than one reminder. My father had, in fact, kept two. His family decided to give one of them to the world.

The green curtain fell. My dad smiled at me. A good soldier and an honorable man who, like thousands of his colleagues, had done his best while facing the unthinkable.

38. Here Lies Honor

2024

It was time to end my journey. But first I needed to revisit a frigid and dark afternoon around November 20, 1944. Major Henley and Colonel Lanham were hunched together at the 22nd Infantry Regimental Headquarters in the Hürtgen Forest.

Maj. Henley: Colonel, sir, I don't care if you break me for it. I meant what I said last night, even though I didn't know it was you on the line. That little patch of woods we're fighting for ain't any good to anybody. No good to the Germans. No good to us. It's the bloodiest damn ground in all Europe, and you make us keep fighting for it. That ain't right.

Col. Lanham: There's nothing in the world that I'd like to do better than tell all you boys to call it off and go home…The only way we can get this thing over is by killing Krauts. To kill them you've got to get to them…I know they've killed lots of our boys in that patch, but we've killed even more of them and that's what counts.[239]

I felt the chill from the cold truth of Charles Wertenbaker's words in their conversation:

> There are some things about war that no explanation can make satisfactory, that no justification can justify.[240]

It had been an honor to get to know the Second Battalion of the proud IVY Division's 22nd Infantry Regiment. I learned how young men, many of them boys really, triumphed. And I could feel some of their anguish as they faltered and sometimes failed. I learned that those who fell with their faces to the foe would always be honored by those who survived them. My father returned to the Chateau de Fontenay to remember. And his story, in turn, led me there.

I had read how great and small acts of courage and compassion mingled freely with acts of cowardice and cruelty during war. How it forced civilians and soldiers to summon their deepest emotional and moral reserves to strive to maintain their humanity as they witnessed the brutal consequences of dehumanizing each other. Sometimes they succeeded and sometimes they failed.

I went and sat on the love seat and looked up at the shelf that was almost empty. My father lay in peace with my mother at Glenwood Cemetery in Darien. His books and papers were in orderly piles in my office.

"Thank you, Dad. If trouble there must be, let me learn from you how to be brave."

That old worry, nervously checking to make sure my heart was beating.

I hoped I would always be strong enough to lead the charge forward. That was what my father never told us about, how he had been one of those facing the foe. I guess it would have sounded like bravado. He had avoided talking about that in his story as well. As Wertenbaker said, in a soldier's war there is occasional humor and heroism but mostly hatred, terror and filth. I was glad Dad had been able to find the humor in being a soldier.

I wished we could sit together, and I could ask him about all the card games and antics he described in his narrative. About what happened to the young men he served with after the war. I could tease him about Beam and Weingart finishing off his good scotch that night, after the shattering battle at the chateau. I wanted to laugh with him and talk about how they all tried to embrace life whenever they could, stubbornly refusing to be overwhelmed by fear of what you had no control over.

"Trust me, if your heart stops beating you will know it."

Although I knew he would be uncomfortable, I wondered if he would let me hug him, and tell him how heartbroken I was to learn that the dashing and engaging Al Mead died in that cruel Hürtgen Forest. I would never mention the unsettling coincidence that both of them were again struck by enemy fire on the same day. But this time Al lost his life while my dad was pulled from the front lines with a hand injury that temporarily rendered him useless as a soldier. I knew this must have been a hard truth for my dad to live with, and I wasn't interested anymore in discussing the why's and how's of war.

I hoped he would have been pleased with the donation of the *Mein Kampf* and my interpretation of his story. That it honored the memory of the men he knew so well who died: Captain Fulton, Captain Newcomb, Lieutenant Cook, Sergeant Wirtzberger, Sergeant Edwards, Sergeant Mead, Private Pierce, Private Carter, and all those lives World War II ended so violently.

I wanted to tell him that courage, for me, now was an infantry soldier waking up in a wretched, frozen foxhole, eating a K-ration, and preparing to move forward knowing, but not thinking about, how today might be the last. It was a parent wrapping their arms around their child and shielding them with their body in life and death as bombs blasted their home. It was the survivor of a concentration camp living out their life—and seeking to not hate, and a soldier forcing himself to remember what he had seen at the camps. It was an infantryman choosing not to kill a surrendering enemy soldier in revenge for the death of a dear buddy. It was a mother whose son had been brutally executed the night before, concealing her agony and quietly serving breakfast to a soldier who was liberating her country.

As for dramatic heroics and strategy, the men I had met helped me see they are not what matters to the infantry soldier. Reliable leadership, camaraderie and survival does. I have begun asking my friends to dust off their parents' contribution to history. The stories I have heard are unexpected and inspiring. I am encouraging them to find the letters home and figure out exactly where and when their family members have served in the military. Then to see if there is a veterans association where they can learn more. I hope these stories will be retold at home to inspire the generations who follow us, not just on Memorial Day and Veterans Day. How ordinary men and women did and continue to do extraordinary things in service to country.

I was satisfied that I understood for myself why soldiers seldom wish to talk about their role in history. War poses questions that can't be answered. As James Davis wrote, "I don't like cemeteries. No one who has ever been to war and returned safely does. Anyone who has ever been in uniform and has lost friends in combat or hears the bugle call of Taps has experienced that same feeling at one time or another."[241]

It had taken 1,600 miles of the soldier's war for me to realize that my dad had summarized his World War II experiences just fine in the stories he told our family. I will always be grateful he kept the journal and his letters for me to discover later. I finally understood why it wasn't his place to tell us much more while he was alive.

His silence honored all those who suffered; all those who died.

I decided that I could not bring my story to its close better than Chaplain Boice had done in his book, *The History of the United States 22nd Infantry Regiment*. I found it in my office then sat back down on the yellow love seat in my living room and opened it. I read out loud, wishing the old wicker chair was still in front of the bookcase.

> On February 20, 1946, Colonel John F. Ruggles inactivated the Twenty-Second Infantry Regiment within the 4th Infantry Division at Camp Burner, North Carolina…Then came the order: "Pass in review…Eyes right" Heads snapped. The generals looked at the soldiers. The soldiers looked at the generals. Neither saw the other but rather the foxholes and hedgerows of Normandy, the crosses at St. Mere Eglise and Henri Chapelle, the match sticks, and the mud of Hurtgen. They saw marching in ranks, in file after file with perfect cadence and deathless spirit, all the men who were not there. Not there? Certainly, they were in the hearts and minds of those who remember, never to forget, in the love of those who would never cease missing them, in the freedom of every American.
>
> And so the men marched off of the parade ground and into the cities and villages and farms, offices or other army posts. And with them went the Twenty-Second United States Infantry Regiment. A dead regiment? Certainly not. Not so long as a single man still lives and remembers. Sleeping, perhaps, but not dead. The Twenty-Second United States Infantry Regiment, the finest Regiment, the beloved Regiment, our Regiment, which gave life to the motto forever etched into our hearts, "Deeds Not Words." [242]

I returned to my mother's words to me on the day my father died. "He was lying prone with his hands folded on his chest. His bloody handprints were on the wall showing how he had fallen."

Five months before his death, I had visited him in the hospital. I remembered my father's ironic chuckle and whispered his calm answer when I asked if he was afraid as he faced the complex cancer surgery, "I haven't been afraid since D-Day."

And then, very slowly, I found what I had been so long in seeking. Politicians move on quickly from a war. Generals move much more slowly. but they too finally leave it behind. A soldier only moves away; in the end, he or she returns to the battlefield to die. I finally understood that although my father achieved standing and wealth during his lifetime, it was a now silent battlefield in Normandy that had informed his life's journey.

I realized that in the last moments of my father's life, as he lay on the floor and folded his hands peacefully on his chest, all the blood and pounding in his head was familiar. He had been there before. I felt certain he had time to offer a silent prayer of thanks for living all those years his fallen peers were denied, to envision his beautiful bride and think of those he loved. And then I saw the familiar Normandy battlefield as he died in the shadow of the chateau. Ellis Phillips' words from the memorial service that long ago day in Florida whispered in the quiet room,

More importantly, I came to appreciate his character and beliefs—a patriotic, unassuming, dependable American—the kind that made our country great and will keep it that way despite national trends and fancies.

Sitting on the love seat, I turned. My father was there beside me. He had been with me all along. I was just too busy looking up.

The Jeep

Epilogue

March 12, 1991

Dear David,

Your letter of 3/9/91 has been forwarded to Stephen Sanders of the 22nd Infantry Regiment Society—also to Tommy Harrison and several others.

    I find it impossible to attend the March 21st memorial service—I would have appreciated the opportunity to see your children—as adults—and to see your mother again after many years...

    Maybe someday I can express to you the respect and affection with which your dad was held by those who knew him—under very trying circumstances.

    My sympathy to you, your mother, and sisters—I too shall miss him very much.

Yours truly,
George R. Bridgeman

When I read this note on March 15, 1991, at my parents' home in Sawgrass, Florida, I did not know George and my father's paths often crossed before and during World War II. I was not aware that my grandparents and his parents were close friends in Galion and Columbus, Ohio.

When he wrote, George Bridgeman lived in Santa Rosa Beach, about a five-hour drive due west from Sawgrass. Had I known what I now know, I would have jumped in my car to meet him. I missed a wonderful opportunity to get to know my grandparents better and to learn about the war years and my dad.

At least now I understand the importance of George's words to my brother. And I can picture Dad and him returning to Belgium in 1944—using their helmets as desks to write to their folks and kidding with each other. Just two young soldiers, answering the call of duty and trying to summon the courage to face what lay ahead.

Honoring that fortitude has crept into my life in small ways. A white azalea now stands at my parent's grave site to honor my sister's battle with cancer. I still see the smile on her face when she welcomed me into her hospital room. Her body, so alive and active a few months earlier, now crippled and failing from metastatic breast cancer. I never heard one word of defeat or complaint during her stand against death. I think of my good friend who is enduring a bone marrow transplant with humor and patience. You can see the suffering in her eyes, but she somehow maintains her enthusiasm for life and for her many volunteer interests. They are heroes—facing the foe.

I returned to the Museum of Jewish Heritage—A Living Memorial of the Holocaust to meet the new curator who helped me with securing images and quotes for my story. She is young and strong. I know that the memory of the girl in the red coat and my father's 22nd Infantry Regiment is in capable hands. This time when I walked through the exhibits, I was struck by the rich history of the Jewish faith and culture, which made the atrocities of the Holocaust even more difficult to comprehend.

My brother now joins me on Memorial Day at our parents' gravesite to honor the men of F Company who died in combat and our sister who met her fate like a soldier. I joined the 4th Infantry Division Association. I am getting to know the men of the 22nd Infantry Regiment Society in the United States and France.

I was welcomed at a recent reunion of the regiment as the daughter of a World War II veteran. The only question I was asked was, "Which battalion did your dad serve under?" I met veterans of Vietnam, the Iraq War, Afghanistan, and the War on Terror. The Library of Congress Veterans History Project has become a new interest. My neighbor who was wounded twice in Vietnam agreed to be my first interview. I still don't agree with most politicians' wars, but that's another matter. I finally understand pride of country and service to it. I am committed to supporting our veterans and active service members. Many of them have also witnessed things of war for which there is no justification.

My journey to find and follow a soldier's footsteps has enriched my life in so many ways. I encourage anyone who enjoyed my story to try walking in the footsteps of someone you love. You may be surprised where the path leads.

"Maj. Henley and the Pups"

"Changing a tire for an audience in Dinkelsbuhl."

"Non-Fraternization"

"40 & 8" Going Home

Acknowledgments

THANK YOU!

John Sisson and Betty Lyons Sisson, my parents, for loving and inspiring me and always being there, even when I wasn't.

Kathryn Marshall, my daughter, for sharing this journey with me at profound moments and giving me a reason for writing the story down.

Paul Lagassé, my unofficial editor, who marked up in red almost every sentence I attempted to craft and patiently listened to my objections as we often reworked sections together.

Ann Bedrick, gifted artist and friend, whose beautiful illustrations of the advance of the 22nd Infantry Regiment from Utah Beach to Bavaria brought the battlefields to life for me.

Ronald Gallimore, Ph.D., friend and scholar, who read the first draft of my memoir, told me to edit it down by 25 percent and signed me up with the Scribblers, my wonderful writing group who offered me the courage and respect to believe I could complete this manuscript.

Rick Taylor, my godbrother and son of my father's best friend Dick Taylor, who read the second draft, had me prune it another 25 percent and offered countless insights and suggestions into the military, life, good writing, and our parents.

Bob Babcock, Historian and publisher, 4th Infantry Division Association, who served with the 22nd Infantry Regiment in Vietnam and knew many of the men who served with my dad. He opened the door.

Richard Breitman, Ph.D., who documents the Holocaust and the atrocities against humanity committed by the Third Reich during World War II through dedicated scholarship on the subject.

My thoughtful and wonderful friends Suzanne Stewart, Rod Rahe, and Kathy and Pete Bennett, who read the early drafts and provided essential observations and suggestions.

The 22nd Infantry Regiment Association, United States and France for tending the history of this proud regiment, and the veterans and active members of the 22nd Infantry Regiment for your courage and commitment to protecting my freedom .

And last, but hardly least, to my husband Hunt Bartine for sharing three years of our marriage with this journey.

About the Author

Karen Marshall lives in Chester County, Pennsylvania, where she worked in historic preservation planning for more than twenty years. One of her proudest accomplishments was the extensive documentation of the Battle of Brandywine (1777) for the American Battlefield Protection Program. Retired, she enjoys sailing with her husband, attending her daughter's choral performances, entertaining her grandchildren, nieces and nephews, and developing projects related to "the stories of the past and how they inform our lives today." She holds a master's degree in Urban Affairs and Public Policy with a certificate in Historic Preservation.

Abbreviations and Slang

Inf. Div.: Infantry Division, commanded by a major general, 10,000–15,000 soldiers.

Bat.: Battalion, commanded by a lieutenant colonel or major, 500–1,000 soldiers in companies (A–M).

Inf. Reg.: Infantry Regiment, commanded by a colonel, 3,000–5,000 soldiers in 3–5 battalions.

Co.: Company, commanded by a captain, 60–250 soldiers in 3–4 platoons.

Platoon: Commanded by a 1st or 2nd lieutenant, 20–60 soldiers in 3–4 squads.

Squad/Section: Commanded by a sergeant, the highest-ranking noncommissioned officer (NCO), 6–10 soldiers.

6X6 trucks: 2 ½ ton supply trucks that formed the backbone of supply transportation.

4 - 22 - 1: 4th Infantry Division–22nd Infantry Regiment–1st Battalion.

1:10,000 maps: U.S. Army Map Service Topographic small-scale maps used for operational planning.

40 & 8: A French box car designed to hold 40 men or 8 horses.

155: 155 mm artillery shell.

57: Bofors 57 mm Automatic Gun.

88's: Shells from a German 88 mm gun, the best overall gun of the war, accurate, lethal and versatile.

A&P: Ammunition and Pioneer, engineers able to perform a range of tasks.

ATS: Auxiliary Territorial Service, women's branch of the British Army

Azimuth: The army uses azimuths to express direction.
BAR: Browning Automatic Rifle.
Bat. C.O.: Battalion Commanding Officer.
Bn. C.O.: Battalion Commanding Officer.
Bn. Hqs.: Battalion Headquarters.
Boche: A contemptuous term for Germans.
CIC: Counter-intelligence Corps.
CID: Criminal Investigation Division.
C.O.: Commanding Officer.
CT 22: 22nd Infantry Regiment Combat Team.
D-Day: Day on which a combat attack or operation is to be initiated.
Duck: DUKW, a six-wheel amphibious truck used to ferry ammunition, equipment and supplies
Exec.: Executive Officer.
FA: Field Artillery.
FFI: French Forces of the Interior.
G-1: General Staff officer for personnel and manpower.
G-2: Senior Army Intelligence Officer.
G.I.: Government Issue, i.e., soldier.
H-Hour: Hour on which a combat attack or operation is to be initiated.
Land Army: Women's Land Army, women working in agriculture for men serving in the military.
LCI: Landing Craft, Infantry, assault vessel that typically carried 200 soldiers.
LCM: Landing Craft, Mechanized, assault vessel that carried vehicles, primarily used in the Allied invasion of North Africa in 1942.
LCT: Landing Craft, assault vessel designed to carry tanks to shore.
LCVP: Landing Craft, Vehicle, Personnel, assault vessel that carried infantrymen or vehicles.
LST: Landing Ship, Tank, ship designed to carry vehicles, cargo and troops to a low beach.

Liaison: Communications facilitator officer.

Looey: Lieutenant

MIS: Military Intelligence Service.

MP: Military Police. SOP: Standard Operating Procedure.

Musette bag: Lighter alternative to full pack used to carry small equipment and personal items.

OD, O.D.: Olive Drab Green (O.G., Olive Green, is the correct term.

P.O.E.: Point of Embarkation

Problems: Specific training exercises for troops.

QM: Quartermaster, the supply officer.

Recon planes: Reconnaissance aircraft.

Red Ball: Red Ball Highway, a truck convoy system from the Normandy beaches to the front lines.

S-1: Personnel Officer.

S-2: Information/Intelligence Officer.

S-3: Operations Officer.

S-4: Logistics Officer.

Shavetail: Derogatory term for a second lieutenant or newly commissioned officer.

SOS: Shit on a Shingle, creamed chipped beef on toast, a pejorative for mess halls.

Val-pac: Garment bag, also referred to as B-4/Valpak.

WAC: Women's Army Corps.

WRENS: Women's Royal Naval Service

Endnotes

1. 4th Infantry Division Association Newsletter, The Ivy Leaves, 2020.
2. Robert O. Babcock, War Stories: Volume II Paris to VE Day (Atlanta: Deeds Publishing, 2019), 140. William Montgomery, Company A, 4th Medical Battalion, 22nd Infantry Regiment, 4th Infantry Division.
3. Charles Christian Wertenbaker, Invasion! (New York: D. Appleton-Century Company, 1944), 144
4. Photo Credit: John F. Sisson and his father, Warren B. Sisson. (from the author's personal collection of unpublished family papers)
5. Unnamed fraternity brother, letter to John Sisson, 1939. (from the author's personal collection of unpublished family papers)
6. John Sisson journal, 1936-1939. (from the author's personal collection of unpublished family papers)
7. Alan Wykes, Himmler (New York, Ballantine Books, Inc.: 1972) Introduction, Barrie Pitt
8. Kaye Ephraim, Desecraters of Memory; Confronting Holocaust Denial. (Jerusalem: Yad Vahsem–International School of Holocaust Studies, 1997), https: jewishvirtuallibrary.org.
9. "Lyons-Sisson Wedding," Columbus Dispatch, December 4, 1948. (from the author's personal collection of unpublished family papers)
10. John Sisson, Letter to William Weingart, February 19, 1980. (from the author's personal collection of unpublished family papers)
11. John Sisson, War Narrative Manuscript, January 1945. (from the author's personal collection of unpublished family papers)
12. Ellis L. Phillips, Letter to Betty L. Sisson, March 5, 1991. (from the

author's personal collection of unpublished family papers)

13. Alan Wykes, Himmler (New York: Ballantine Books, Inc., 1972) Introduction

14. Wikipedia, https://commons.wikimedia.org/wiki/File:Bundesarchiv_Bild_152-11-12,_Dachau,_Konzentrationslager,_Besuch_Himmlers.jpg This file is licensed under the Creative Commons Attribution-Share Alike 3.0 Germany license. Attribution: Bundesarchiv, Bild 152-11-12 / CC-BY-SA 3.0

15. Alan Wykes, Himmler (New York: Ballantine Books, Inc., 1972), 6.

16. Jim Fitzgerald, "Himmler's Annotated Hitler," Washington Post, December 8, 1998.

17. Wykes, Himmler, 112.

18. The Museum of Jewish Heritage—A Living Memorial to the Holocaust, Mission, https://mjhnyc.org.

19. Fred Somers letter to Betty Sisson, March 10, 1991 (from the author's personal collection of unpublished family papers)

20. Richard Breitman, The Architect of Genocide, Himmler and the Final Solution (Hanover and London: Brandeis University Press of New England, 1991), 4.

21. Ibid., 219.

22. Ibid., 219

23. Ibid., 219

24. Ibid., 220

25. Richard Breitman, Mein Kampf and the Himmler Family: Two Generations React to Hitler's Ideas, Holocaust and Genocide Studies, Volume 13, Number 1 (Spring 1999), 90

26. Breitman, Architect of Genocide, 3.

27. "You Couldn't Grasp It All; American Forces Enter Buchenwald," The National WWII Museum (April 9, 2021), www.nationalww2museum.org.

28. Breitman, Architect of Genocide, 4.

29. Robert McFadden, "Obituary Robert M. Morgenthau," New York

Times, July 21, 2019.

30. Betty Sisson letter to Robert M. Morgenthau. (from the author's personal collection of unpublished family papers)

31. Charles Wertenbaker, Invasion (New York: D. Appleton-Century, 1944), 1.

32. S.Sgt. Albert R. Simpson, Arlington, Virginia, October 21, 1967 National Archives and Records Administration. This file is a work of a U.S. Army soldier or employee, taken or made as part of that person's official duties. As a work of the U.S. federal government, it is in the public domain in the United States. Image cropped.

33. Graham Nash, Chicago (Songs for Beginners, 1971, Atlantic Records)

34. Cornelius Ryan, The Longest Day (New York, Simon and Shuster, 1959), Foreward

35. Charles Christian Wertenbaker, Invasion! (New York: D. Appleton-Century Company, 1944), Dust Jacket

36. Unsourced newspaper clipping, Kathryn Phillips Diaries, edited by Noel Phillips. (from the author's personal collection of unpublished family papers)

37. https://dod.defense.gov/Portals/1/features/2016/0516_dday/docs/d-day-fact-sheet-the-beaches.pdf. "D-Day: The Beaches," PDF.

38. Cornelius Ryan, The Longest Day (New York: Simon & Shuster, 1959), 99.

39. Omar N. Bradley, A Soldier's Story (New York: Henry Holt and Company, 1951), 564.

40. Huey Freeman, "Book recalls 4th Infantry's Valor on Utah Beach," Herald and Review, Decatur, Illinois (April 8, 2008).

41. Stephen Ambrose, D-Day June 6, 1944: The Climactic Battle of World War II (New York: Simon & Shuster, 1994), 292-293.

42. G.H. Bennett, Destination Normandy: Three Regiments at D-Day (Praeger Security International: 1967), xvi.

43. John Sisson, WW II Narrative, (from the author's personal collection of unpublished family papers)

44. John Sisson journal, 1936-1939. (from the author's personal collection of unpublished family papers)

45. The quote is attributed to popular culture in Britain, 1942-1945.

46. Western Union, Kathryn Sisson Phillips to her brother Warren and Margaret Sisson, John Sisson's parents, Galion, Ohio, June 6, 1944. (from the author's personal collection of unpublished family papers)

47. Western Union, The Adjutant General to Warren and Margaret Sisson, Galion, Ohio, June 30, 1944. (from the author's personal collection of unpublished family papers)

48. James K. Cullen, Band of Strangers (Jersey City, Nazzaro & Price Publishing, 2018), Acknowledgements.

49. Karen Marshall, July 2019. (from the author's personal collection of unpublished family papers).

50. Charles Wertenbaker, Invasion! (New York: D. Appleton-Century Company: 1944), 108.

51. Robert Babcock, War Stories, Volume I (Atlanta: Deeds Publishing: 2019), 160

52. Wertenbaker, Invasion, 99.

53. Dr. William S. Boice, Chaplain, History of the Twenty-Second United States Infantry in World War II (Unpublished: 1959), 24-25. (from the author's personal collection of unpublished family papers) (This manuscript has recently been published by Deeds Publishing, 2024.)

54. Wertenbaker, Invasion, 147 and 154.

55. Ibid., 75

56. Atkinson, Rick, The Guns at Last Light, The War in Western Europe, 1944-1945 (New York, Henry Holt & Company: 2013), 112.

57. Blumenson, Martin, U.S. Army History: Break Out and Pursuit (Washington, U.S. Government Printing Office,1961), 176.

58. John Sisson, Letter to Senator Daniel Inouye, July 12, 1987. (from the author's personal collection of unpublished family papers)

59. J.Q. Lynd," Chateau de Fontenay Report," (June 9-10, 1944). (from the

author's personal collection of unpublished family papers)

60. Robert Rush, Hell in the Hurtgen Forest. (Lawrence: University Press of Kansas, 2001), Forward xiii

61. Cornelius Ryan, The Longest Day (New York: Simon and Shuster, 1959), 287.

62. Normandy Then and Now, "Batterie du Holdie," www.normandythenandnow.com

63. Rick Atkinson, The Guns at Last Light, The War in Western Europe, 1944-1945 (New York: Henry Holt & Company, 2013), 61.

64. John Sisson letter to Richard I. Taylor, July 14, 1944. (from the author's personal collection of unpublished family papers)

65. John Sisson letter to Lt. General Glenn D. Walker, May 22, 1989. (from the author's personal collection of unpublished family papers)

66. John Sisson letter to Richard I. Taylor, July 8, 1944. (from the author's personal collection of unpublished family papers)

67. Charles Wertenbaker, Invasion! (New York, D. Appleton-Century Company, 1944), 65.

68. Huey Freeman, "Book recalls 4th Infantry's valor on Utah Beach," Herald and Review, Decatur, Illinois, April 8, 2008

69. Dr. William S. Boice, History of the Twenty-Second United States Infantry in World War II (Unpublished: 1959),17. (from the author's personal collection of unpublished family papers)

70. Ibid., 18.

71. John Sisson letter to William (Bill) Weingart. February 19, 1980. (from the author's personal collection of unpublished family papers)

72. Bill Weingart letter to John Sisson, July 1, 1980. (from the author's personal collection of unpublished family papers)

73. U.S. Army History, Utah Beach to Cherbourg: 6-27 June 1944. (War Department's Historical Division: 1948) (reprinted Washington, Center of Military History United States Army, 1990), 107.

74. Boice, History of the Twenty-Second Infantry, 21. (from the author's

personal collection of unpublished family papers)

75. Dr. William S. Boice, History of the Twenty-Second United States Infantry in World War II (Unpublished: 1959), 22. (from the author's personal collection of unpublished family papers)

76. John Sisson letter to Bill Weingart, February 19, 1980. (from the author's personal collection of unpublished family papers)

77. John Sisson letter to Lt. General Glenn D. Walker, May 22, 1989. (from the author's personal collection of unpublished family papers)

78. Robert Babcock, Volume I. (Atlanta, Deeds Publishing, 2019), 163.

79. Bill Weingart letter to John Sisson, July 1, 1980. (from the author's personal collection of unpublished family papers)

80. Boice, History of the Twenty-Second Infantry, 21.

81. Matthew A. Rozell, The Things Our Fathers Saw (New York: Woodchuck Hollow Press, 1961), 343.

82. Karen Marshall, Museum of Jewish Heritage – A Living Memorial to the Holocaust in 2023. Core exhibition, "The Holocaust: What Hate Can Do." (from the author's personal collection of un-published family papers)

83. Bill Weingart, letter to John F. Sisson, July 1, 1980. (from the author's personal collection of unpublished family papers)

84. The Great Courses, World War II: A Military and Social History, Professor Thomas Childers, University of Pennsylvania. Published Great Courses/teaching company, January 1, 1998.

85. The National WWII Museum Profile, An Architect of Terror: Heinrich Himmler and the Holocaust, www.nationalww2museum.org.

86. Museum of Jewish Heritage – A Living Memorial to the Holocaust, Traveling Exhibit, Auschwitz, Not Long Ago, Not Far Away, May 8, 2019-May 3, 2021.

87. Museum of Jewish Heritage – A Living Memorial to the Holocaust in 2023 in their core exhibition "The Holocaust: What Hate Can Do."

88. The Balzec Killing Center, described by SS-Obersturmfuhrer (Second Lieutenant) Kurt Gerstein, 1945, Auschwitz, Not Long Ago, Not Far Away,

May 8, 2019-May 3, 2021, Museum of Jewish Heritage – A Living Memorial to the Holocaust.

89. Charlotte Delbo, Auschwitz and After, Second Edition (New Haven: Yale University Press, 1995.) Delbo was incarcerated in Auschwitz in January 1943.

90. Richard Breitman, The Architect of Genocide, Himmler and the Final Solution (Hanover and London: Brandeis University Press of New England, 1991), 250.

91. Marc Romanych, "Third Reich in Ruins, US National Archives, RG 111SC-207089." Caption: Himmler's wife Marga (Margarete) and daughter Gudrun lived here during the Third Reich period. www.thirdreichruins.com/tegernsee.htm

There are now multiple stories on the web about the Himmler home since it was purchased and as of the fall of 2023 converted into a hotel/restaurant. The development has been surrounded by criticism even though the owner hopes to make it a center of peace and tolerance. One critic noted: "What next, should we build a rollercoaster ride in Dachau?" Rob Hyde, "The JC, Nazi SS chief Heinrich Himmler's Bavarian lakeside villa transformed into hotel," (September 1, 2023).

92. Earl West letter to John Sisson September 5, 1975. (from the author's personal collection of unpublished family papers)

93. John Sisson letter to Bill Boice April 3, 1980. (from the author's personal collection of un-published family papers)

94. George Wilson letter to John Sisson, October 26, 1990. (from the author's personal col-lection of unpublished family papers)

95. John Sisson letter to George Wilson, November 2, 1990. (from the author's personal collection of unpublished family papers)

96. John Ruggles note to John Sisson, June 1, 1982. (from the author's personal collection of unpublished family papers)

97. John Sisson letter to General Ruggles, July 1, 1982. (from `the author's personal collection of unpublished family papers)

98. Laurence Binyon, For the Fallen, (London: The Times, September 1914).
99. The framed map is in the private collection of Albert F. Goetze, Jr. The Binyon poem is modified and emphasized as part of the memorial.
100. "A Day at the Beach," Herald and Review, Decatur, Illinois, 2003.
101. Binyon, For the Fallen.
102. James Cullen, Band of Strangers (Jersey City, Nazzaro & Price Publishing, 2018), 285.
103. Ibid., 297-298.
104. Ibid., 298
105. "Elliott Richardson addressing the 1973 22nd Infantry Society Reunion," 22nd Infantry Association Newsletter, 1973. (from the author's personal collection of unpublished family papers)
106. Bob Babcock email to Karen Marshall. Robert Babcock is President Emeritus and Historian, National 4th Infantry Division Association. (from the author's personal collection of unpublished family papers)
107. "A Day at the Beach," Herald and Review, Decatur, Illinois, 2003.
108. John Growth, Studio Europe (New York: The Vanguard Press, 1945), 206.
109. Omar N. Bradley, A Soldier's Story (New York, Henry Holt and Company, 1951), 321.
110. John Sisson's 1945, scrapbook. (from the author's personal collection of unpublished family papers)
111. Lt. Wilson, 7-9.
112. Sgt. Cullen, 9.
113. Boice, History of the Twenty-Second Infantry Regiment, 26.
114. Ibid, 26.
115. Ibid, 26.
116. Ibid, 26.
117. Ibid, 27.
118. Ibid 27. First Sergeant Kenyon was killed in action.
119. Ibid, 28.

120. Lt. Wilson, 7.
121. Martin Blumenson, Breakout and Pursuit (Washington: U.S. Government Printing Office, 1961), 179.
122. Lt. Wilson, 13-14.
123. Boice, History of the Twenty-Second Infantry Regiment, 30.
124. Lt. Wilson, 17.
125. Lt. Wilson, 43-45.
126. Sgt. Cullen, 56.
127. Sgt. Cullen, 79.
128. Gen. Bradley, 407.
129. Lt. Wilson, 58.
130. Gen. Bradley, 392.
131. Robert O. Babcock, War Stories: Volume II Paris to VE Day. (Atlanta: Deeds Publishing, 2019), 251-252.
132. Ibid., 260-261.
133. Ibid, 262-263.
134. Boice, History of the Twenty-Second Infantry Regiment, 41.
135. Lt. Wilson, 65.
136. Boice, Twenty-Second Infantry Regiment, 47.
137. Ibid, 73.
138. Lt. Wilson, 73.
139. Boice, History of the Twenty-Second Infantry Regiment, 50.
140. Sgt. Cullen, 144.
141. Boice, Twenty-Second Infantry Regiment, 48.
142. Charles Whiting, "Ernest Hemingway and the Ivy Leaguers in World War II," www.warfarehisotrynetwork.com.
143. Sgt. Cullen, 166.
144. Ibid, 171-173.
145. Lt. Wilson, 85.
146. Ibid., 91.
147. John Groth, Studio Europe (New York, 1945), 208.

148. Boice, Twenty-Second Infantry Regiment, 56.
149. Groth, Studio Europe, 220-221.
150. George Bridgman letter home, 10/6/1944. (from the author's personal collection of unpublished family papers)
151. Lt. Wilson, 120.
152. Boice, Twenty-Second Infantry Regiment, 56.
153. Lt. Wilson, 129.
154. Philip Rahy, "Across the River and Into the Trees, by Ernest Hemmingway," Commentary, October 1950 Literature.
155. Sgt. Cullen, 405-406.
156. Boice, Twenty-Second Infantry Regiment, 56.
157. Gen. Bradley, 442.
158. Gen. James Gavin, "Bloody Huertgen: The Battle That Should Never Have Been Fought," American Heritage (December 1979, Volume 31, Issue 1).
159. Robert S. Sterling Rush, Hell in the Hurtgen Forest, The Ordeal & Triumph of an American Infantry Regiment (Lawrence: University of Press Kansas, 2001), 281 & 182.
160. Ibid., 3.
161. Ibid., 348.
162. Boice, History of the Twenty-Second Infantry Regiment, 57.
163. Ibid., p. 65.
164. Robert Babcock, War Stories Volume II (Atlanta: Deeds Publishing, 2019), 102.
165. Babcock, War Stories II, 67-71.
166. Lt. Wilson, 151.
167. Ibid., 168.
168. Rush, Hell in the Hurtgen Forest, 247.
169. Lt. Wilson., 173.
170. Boice, Twenty-Second Infantry Regiment, 99.
171. Babcock, War Stories II, 96.

172. Rush, Hell in the Hurtgen Forest, 272.
173. Boice, Twenty-Second Infantry Regiment, 102. In 2002, Robert Rush cited the casualties at 2,805 (86%) out of the normal complement of 3,253 officers and men.
174. Ibid., 103.
175. John Groth, Studio Europe (New York, 1945), 254.
176. Ibid., 255.
177. Boice, Twenty-Second Infantry Regiment, 65.
178. National 4th Infantry (IVY) Division Association, www.4thinfantry.org/about-us/division-history.
179. Boice, History of the Twenty-Second Infantry History, 109.
180. Robert Babcock, War Stories Volume II (Atlanta: Deeds Publishing, 2019), 112.
181. Lt. Wilson, 197.
182. Boice Twenty-Second Infantry History, 117.
183. Ibid., 117.
184. Ibid., 118.
185. Ibid., 118.
186. Lt. Wilson, 201.
187. Boice, Twenty-Second Infantry History, 119.
188. Lt. Wilson, 202.
189. Boice, Twenty-Second Infantry History, 120.
190. Lt. Wilson, 206.
191. Boice, Twenty-Second Infantry History, 120.
192. Lt. Wilson, 212.
193. Boice, Twenty-Second Infantry History, 120.
194. Ibid., p. 121.
195. Robert Babcock, After Action, December 31, 1944, December 1944 Casualties.
196. Boice, Twenty-Second Infantry History, 122-123.
197. Lt. Wilson, 220.

198. Boice, History of the Twenty-Second Infantry, 126.
199. Sgt. Cullen, 254, 262.
200. Ibid., 261-262.
201. Ibid., 312-314.
202. Ibid., 169.
203. Lt. Wilson, 232-233.
204. Ibid, 234-235.
205. Boice, History of the Twenty-Second Infantry, 135.
206. Lt. Wilson, 248.
207. Boice, History of the Twenty-Second Infantry, 136.
208. James Martin Davis, "Memorial Day Our Nation's Time to Remember," Omaha World-Herald (Omaha, Walsworth Publishing, 2021) Nov. 8, 1981. Omaha attorney and Vietnam vet, James Martin Davis, wrote a Memorial Day essay for four decades for The World-Herald. Davis died in 2021. His first essay, "A Soldier Remembered: I am Proud of the Uncle I Never Knew" was printed Nov. 1, 1981, 20-21.
209. Ibid., 22.
210. Robert Rush, Hell in the Hurtgen Forest (University of Press Kansas, 2001), 309.
211. Lt. Wilson, 229.
212. 4th Infantry Division, 22nd Infantry Regiment "Souvenir Yearbook" and Roster. (Harold W. Blakeley, Major General, United States Army, Commanding, 1945), 91-92.
213. Robert O. Babcock, War Stories: Volume II Paris to VE Day. (Atlanta: Deeds Publishing, 2019), 208.
214. Boice, History of the Twenty-Second Infantry, 140.
215. Ibid., 142.
216. Ibid., 143.
217. Ibid., 144.
218. Ibid., 144.
219. Ibid., 140.

220. Babcock, War Stories: Volume II, 172-173.
221. Ibid, 206.
222. Boice, History of the Twenty-Second Infantry, 146.
223. Ibid., 147.
224. Ibid., 147.
225. Babcock, War Stories: Volume II, 209.
226. Boice, History of the Twenty-Second Infantry, 151.
227. Babcock, War Stories: Volume II, 173-174.
228. Ibid, 209.
229. Matthew A. Rozell, The Things Our Fathers Saw (New York, Woodchuck Hollow Press, 1961, 224-226.
230. Boice, History of the Twenty-Second Infantry, 152-153.
231. Ibid., 154.
232. After Action: Casualties – From 6 June 1944 to 9 May 1945.
233. Walk of Honor, Combat Infantrymen's Association, Dedicated 2011 National Infantryman Museum, Columbus, Georgia.
234. Dr. William S. Boice, History of the Twenty-Second United States Infantry in World War II (Unpublished: 1959),152. (from the author's personal collection of unpublished family papers)
235. Ibid, 152.
236. Evaluation Report No. 243, "Based on Intelligence Report EW-Hc 107," May 30, 1945. August 5, 1945. Library of Congress.
237. Omar N. Bradley, A Soldier's Story (New York: Henry Holt and Company, 1951), 314.
238. Matthew A. Rozell, The Things Our Fathers Saw (New York, Woodchuck Hollow Press, 1961), 316-317.
239. Dr. William S. Boice, History of the Twenty-Second United States Infantry in World War II (Unpublished: 1959), 93. (from the author's personal collection of unpublished family papers)
240. Charles Wertenbaker, Invasion (New York, D. Appleton-Century, 1944) p. 100.

241. James Martin Davis, "Memorial Day Our Nation's Time to Remember," Omaha World Herald (Omaha, Walsworth Publishing, 2021) p. 14.
242. Boice, History of the Twenty-Second Infantry, p. 163.

NOTE

The endnotes 111-232 were referenced as regular notes with the following exceptions:

- 4th Infantry Division, 22nd Infantry Regiment "Souvenir Yearbook" and Roster. (Harold W. Blakeley, Major General, United States Army, Commanding, 1945) Referenced by 22nd Infantry Regiment Roster and page number.
- After Action Reports 4th Infantry Division. 4th Infantry Division Association, www.4thinfantry.org. Referenced by After Action and the date of the entry.
- Dr. William S. Boice, History of the Twenty-Second United States Infantry in World War II (Unpublished: 1959) referenced by Boice, History of the Twenty-Second Infantry Regiment.
- Gen. Bradley, referenced by page number from A Soldier's Story (New York, Henry Holt and Company, 1951)
- Pvt. Cassar Diary, referenced by the date of the entry. Permission by family permission.
- Sgt. Cullen, referenced by page number from Band of Strangers (Jersey City, Nazzaro & Price Publishing, 2018)
- Capt./Maj. Henley, Unpublished Diary, referenced by the date of entry.
- Lt. Sisson, referenced by the date of entry. Letters and excerpts from Part 4: One Soldier's Story. (from the author's personal collection of unpublished family papers).

- Lt. Wilson, referenced by page number from If You Survive (New York, Ballantine Books, 1987).

PHOTO CREDITS BY CHAPTER:

Cover Photo: Kathryn Marshall, 2019
Chapter 24: John Sisson, 1979; Karen Marshall, 2019
Chapter 25 Photo Credits: John Sisson, 1979
Chapter 31: 22nd Infantry Regiment Roster, 60 and 34.
Chapter 32: 22nd Infantry Regiments Roster, 103.
Chapter 33: File:SC 270663, (49346150276).jpg Wikimedia Commons, US National Archives, 8th Infantry Regiment, 4th Infantry Division and 22nd Infantry Regiment Roster, 105.
Chapter 35: 22nd Infantry Regiment Roster, 41, 43
Chapter 36: John Sisson 1945 Scrapbook. 22nd Infantry Regiment Roster, 72.
Epilogue: John Sisson Scrapbook, Circa summer 1945, Dinklesbuhl, Germany. (from the author's personal collection of unpublished family papers)

Printed in the USA
CPSIA information can be obtained
at www.ICGtesting.com
CBHW030520230724
12002CB00002B/5